M A T H

FOR THE PEOPLE

Basic math literacy

ARITHMETIC ▲ FRACTIONS ▼ DECIMALS ▲ PERCENTS

Mark Charalambous

Math for the People: Basic Math Literacy

First Edition

Copyright © 2016 by Arjuna's Arrow Publishing. All rights reserved.

ISBN-13: 978-0692663158

ISBN-10: 0692663150

Contents

Foreword

I began writing this book for a non-credit community college class that never materialized. The administrators thought there might be a need for a short 4- or 5-session evening course for adults in the community who for one reason or another never learned basic math.

I imagined the target audience to be a blue-collar shop worker who had learned the math he needed specifically for his job, but desired a better understanding of fractions, decimals and/or percents either for a promotion or just his own edification. Or, perhaps, an empty-nester working in retail who always felt embarrassed by her weaknesses at the cash register and decided now was the time to do something about it.

As is often the case, it won't surprise me if it turns out I have seriously misjudged the audience for this book. We know that math illiteracy (jargon alert: "innumeracy") is all too common. But for some reason, unlike language illiteracy, many such people afflicted don't seem to be ashamed to admit they can't do math. How often do we hear, "I suck at math"? It certainly doesn't bear the same stigma as illiteracy, which is usually treated as a closely guarded family secret. Sometimes it almost seems like "sucking at math" is in some perverse way a badge of pride, as though it's "cool."

Regardless, I wrote this book for the person who never "got" *some* aspect of basic math—perhaps subtraction... or percents... but most likely fractions. A reader with *some* basic math knowledge but wants to bone up on, say, adding fractions, may prefer to go directly to one of the step-by-step examples in that chapter. Once there, they may find themselves hunting around for some prior definition or explanation, or proceed to any of the You-Try-It examples or the exercises at the end of the section.

For purposes of this book I define *basic* math to include: a casual overview of numbers, whole-number arithmetic, fractions, decimals, and percents. I avoid signed numbers, other than to mention in the first short chapter on numbers that they are part of the integers, and after that I am ignoring them. *Math for the People* is an ideal textbook choice for any course that includes these topics.

I attempt to write in a light, conversational tone that will engage the average adult reader. The book is designed as an all-in-one tool—all that's required is a pencil. Portable. No need for internet access. No bells and whistles. Nothing to intimidate the person who just wants some quiet time to fulfill what is perhaps a bucket-list goal: learning to communicate with the world of numbers. It's intended as a tutorial on everyday math, the math that is occasionally needed for everyday life tasks, usable math for the everyday man or woman. Hence: *Math for the People*.

Chapter 1   Numbers

Math begins with numbers. We can use our imaginations to speculate on how the various number systems we use developed over time. When we were living in caves the only thing we needed math for was to count things: rocks, children, wives, etc. So we would create a primitive system of counting numbers: 1,2,3 and so on. The concept of a number to represent nothing, i.e., the number zero, would probably not occur to our most ancient of ancestors.

Natural or Counting numbers

Indeed this primary system of numbers, 1,2,3, ... is referred to as the *Counting* numbers, though it has a more formal name: the *Natural* numbers. The three dots "..." (called an *ellipsis*) means the pattern continues forever.

Natural or Counting numbers: 1, 2, 3, 4, . . .

Whole numbers

The concept of a number to represent nothing is actually a relatively sophisticated concept. We are familiar with the Roman numerals, I,V, X, L, C, D, M, such as we see at the end of the movie credits to indicate the year the film was made, but it doesn't occur to most people that this system did not include a numeral for zero. The Roman civilization was technologically advanced. Besides roads and multi-story buildings, they had aqueducts and indoor plumbing (well, the very rich did; the common folk had to use public water facilities). In fact, their building technology required the knowledge of the *Irrational* numbers, such as π (the Greek letter Pi)—which we will mention in a while—yet they got by without zero!

The history of zero may be a fascinating topic, but it is not what you are here for. We just need to note that when we add the number 0 to the Natural/Counting numbers, the new set of numbers is called the whole numbers.

Whole numbers: 0, 1, 2, 3, 4, . . .

Returning to our friend living in the cave with his family and his rocks, at some point he joined with other people and formed a larger community. When social relationships formed within a community, economics cannot be far behind. At some point we can imagine that these primitive folks would find a need to create a math system to do more than just count. We can well imagine that the first operation to be created would be addition.

Assuming that the only numbers used were the Natural numbers, if we think about addition with this set of numbers we soon realize that there's something very neat about adding with the Natural numbers: No matter how many or which numbers we add, we will always wind up with another Natural number.

Integers

Once a need for addition has been found, subtraction would inevitably follow. For, if we add 2 to 3 to get 5, we will at some point want to know what happens if we attempt to reverse this process—that is, if we start with 5 and end up with 3, what happened in between?

When our primitive ancestor mathematicians played with this new concept it wouldn't be long before they stumbled on a serious problem. As long as they subtracted a smaller number from a larger one, everything was cool. It was just like addition. You always got another Natural number. But if you subtracted a number from itself, there was nothing left! This by itself might lead directly to the creation of zero. But an even more serious calamity ensued if they tried to subtract a larger number from a smaller one. Such a subtraction assuredly put them outside their "mathematical universe." A politician might say it represented an "existential threat" to their world and demand the exile of the heretic who discovered it!

In fact, the operation of subtraction on the Natural numbers requires a new larger set of numbers to explain what happens when a larger number is subtracted from a smaller. Such a subtraction produces what we call a *negative number*. For example, $4 - 6 = -2$. In this book, we are focusing on math basics, and for our purposes we are not including negative numbers. However, the larger set of numbers that includes the negative numbers is called the *Integers*.

Integers: . . . -4, -3, -2, -1, 0, 1, 2, 3, 4, . . .

The integers proved to be an enormously useful set of numbers that contributed much to human progress, but they too, eventually became insufficient.

Rational numbers

Continuing with our thought experiment regarding the development of the arithmetic operations, we imagine how the need for a new operation would arise when there is what we call *repeated addition*, for example, if we wished to add $3 + 3 + 3 + 3 + 3 + 3 + 3 + 3$. We might develop a table to remind us that the sum of eight 3's is 24. This could be quite a timesaver. Hence: *multiplication*.

Ignoring the negative numbers, we soon realize that multiplication behaves just like addition with the Natural numbers: if we multiply any Natural number, we get another Natural number. 3 X 8 = 24. This is nice! We like this. We don't exile the man that comes up with this, we reward him with a new set of rocks!

However, similar to our discovery of subtraction from seeking to reverse the process of addition, we wonder what happens if we reverse multiplication; that is, if we start with 24 and end up with 3, what happened in between?

Hence: *division*.

Playing with this new mathematical toy we *very* quickly run into a new problem. Having learned our lesson from what happened to the guy who tried to subtract a larger number from a smaller, we may stipulate from the get-go than we're *only* going to use division to divide a larger number, like 24, by a smaller number, like 3 or 8. As long as we stay with numbers formed from our multiplication tables, everything is okay. But as soon as we stray—say, by attempting to divide 24 by 5—we run into another problem. We don't get a Natural number! At this point we may decide to examine our decision not to use division on smaller numbers by larger ones. Perhaps these issues are related?

Fractions

After casting out another tribe member for mucking up our world, we eventually face up to the reality: yet another system of numbers is required. The numbers created by division that do not resolve to a Natural number are called *fractions*. Essentially, what we do is just acknowledge that dividing 24 by 5, or 3 by 24, is a valid concept, and we choose a way to represent it with numerals and a new symbol, the fraction bar, like this: $\frac{3}{24}$, and $\frac{24}{5}$.

The formal name for the set of numbers that includes fractions along with the Integers is called the *Rational* numbers.

Unlike the previous sets of numbers, we cannot represent the Rationals as a comma-separated list of numbers in sequence, because between any two fractions, no matter how close in size, you can always find another one. Of course, it turns out that between any two fractions, such as ½ and ⅓ for example, there exists an *infinite* set of numbers. We define the rational numbers by describing how we create them: by dividing an integer by another integer.

Rational numbers: All numbers that can be represented by the division of one integer by another[1].

There is another set of numbers, the *Irrational* numbers, which are needed to complete the set of numbers that is used for most worldly needs, the *Real* numbers. The Irrational numbers include numbers like π, a number necessary to describe measurements involving a circle (actually any cyclically repeating behavior), as well as numbers like $\sqrt{2}$ (the *square root* of 2).

Our discussion on numbers ends here. For our purposes the numbers that we will use will be the whole numbers (Natural/Counting numbers plus zero) and the (non-negative) Rational numbers (fractions) ***only***. In this book, we will learn how to perform the four arithmetic operations on these numbers, including the decimal representations of the Rationals, as well as numbers written in percent form.

In Chapter 2, we begin by refamiliarizing ourselves with the four arithmetic operations, addition, subtraction, multiplication and division, confined to the whole numbers only.

---

[1] Except division by zero, which is not defined.

Chapter 2 Arithmetic

In order to understand how arithmetic works we have to understand our *place-value* number system.

2.1 Understanding place-value

Our number system is a *decimal*, or base 10, system. This means that there must be ten different numerals (aka *digits*): 0, 1, 2, 3, 4, 5, 6, 7, 8, 9. All numbers can be made up of these numerals in various combinations.

Each position, called a *place value*, can include any of the 10 numerals.

For example, in the number 3 0 9 5 each place has a value assigned to it.

Units or Ones place

Tens place

Hundreds place

Thousands place

So, the number 3095 means

$$
\begin{array}{rrcr}
 & 5 \times 1 & = & 5 \\
+ & 9 \times 10 & = & 90 \\
+ & 0 \times 100 & = & 0 \\
+ & 3 \times 1000 & = & 3000 \\
\hline
 & \text{Total} & & 3095 \\
\end{array}
$$

Note when writing numbers greater than 999 in standard notation, we write commas to separate the digits into groups of 3, called *periods*. The periods are *ones*, *thousands*, *millions*, *billions*, and *trillions*. The periods are used to write the commas in big numbers and also when we write numbers in words, as we do when we write a check.

In standard notation, we write as 20,746—*not 20746*. We write 2,067,087—*not 2067087*, etc.

2.2 Addition

There is a tried-and-true method of adding numbers that has been in use for many generations. Chances are, the reason you're reading this book is not to learn how to add—however, it's entirely possible that this is exactly why some people are reading this book. I make no assumptions about what you know, which is why after the brief overview on numbers in the previous chapter, the very next thing we are going to explain is how to add whole numbers in the tried-and-true method, which I refer to as the *Standard Addition algorithm*. An *algorithm* is a series of steps, repeated as necessary to solve a problem.

Following that, we will provide a more in-depth explanation of what is going on in the Standard Addition algorithm, which you may want to skip if all you're looking for at this time is just 'How to do it.' I will also explain a method for mentally adding numbers that is consistent with the federal Common Core standards presently being implemented in grade school.

We use the words *sum* and *total* for the answer to an addition problem. The numbers being added are called *addends*.

2.2.1 The Standard Addition algorithm

To add two or more whole numbers we right-align (or "right-justify") them. To add 27, 63, and 152, we align the numbers vertically and, usually, write a "+" at the left of the bottom number to indicate we are doing an addition problem, like this:

$$
\begin{array}{r}
2\ 7 \\
6\ 3 \\
+\ 1\ 5\ 2 \\
\hline
\end{array}
$$

We will compute and then write the sum in a row underneath the line. We add the numbers column by column, starting at the right-most column (the *ones* column). In the first of the columns, the numbers 7, 3 and 2 add to 12. When a sum is greater than 9, as is the case here, we write only the last digit (2 in this case), and "carry" the other digit (1, in this case). When we carry a digit we write it at the top of the next column over, as shown here:

```
      1
      2  7
      6  3
   1  5  2
   _____
         2
```

(Note when we carry we usually write the digit smaller and a little to the left to differentiate it from the actual numbers in the addition problem.)

Now when we add the digits in the next column (*tens* place, second from the right) we include the carry digit (1 in this case); so we have to add 1, 2, 6 and 5, which sums to 14. We put the 4 in the total row and carry the 1 onto the next column over:

```
   1  1
      2  7
      6  3
   1  5  2
   _____
      4  2
```

There is only one column of numbers remaining in the *hundreds* place, it has the carry digit 1, and the digit 1 from 152. Adding 1 and 1 we usually get 2, which we write in the total row, and now the problem is finished.

```
   1  1
      2  7
      6  3
   1  5  2
   _____
   2  4  2
```

The sum is 242.

Let's do one more example, and then you can try some.

Example. Add 851, 27, 974, and 86

Step 1. Vertically right-align the numbers.	$\begin{array}{cccc} & 8 & 5 & 1 \\ & & 2 & 7 \\ & 9 & 7 & 4 \\ + & & 8 & 6 \\ \hline \end{array}$
Step 2. Adding the digits in the ones column: 1, 7, 4, and 6, gives 18. Write 8 in the ones place in the total row and carry the 1.	$\begin{array}{cccc} & & 1 & \\ & 8 & 5 & 1 \\ & & 2 & 7 \\ & 9 & 7 & 4 \\ + & & 8 & 6 \\ \hline & & & 8 \end{array}$
Step 3. Add the digits in the tens column: 5, 2, 7, 8 and 1, the carry digit. The sum of these is 23. Write 3 in the tens place in the total row and carry the 2.	$\begin{array}{cccc} & 2 & 1 & \\ & 8 & 5 & 1 \\ & & 2 & 7 \\ & 9 & 7 & 4 \\ + & & 8 & 6 \\ \hline & & 3 & 8 \end{array}$
Step 4. Add the digits in the last column (the hundreds column): 8, 9, and 2, the carry digit, which total to 19. Since there are no more numbers to add, write 19 to the left of 38 in the total row. The sum is 1,938.	$\begin{array}{cccc} & 2 & 1 & \\ & 8 & 5 & 1 \\ & & 2 & 7 \\ & 9 & 7 & 4 \\ + & & 8 & 6 \\ \hline 1 & 9 & 3 & 8 \end{array}$

You-Try-It 1

Add:

a) 45 plus 267

b) 84, 104, and 79

c) 612 + 814 + 63 + 192

d)

$$
\begin{array}{r}
4\ \ 2\ \ 0 \\
9\ \ 7 \\
8\ \ 9\ \ 3 \\
+\qquad 6\ \ 5 \\
\hline
\end{array}
$$

2.2.2 A deeper look: Addition by decomposition

In our example where we added 851, 27, 974, and 86, when we added the digits in the ones place $(1 + 7 + 4 + 6)$ we got 18. Eighteen is 10 plus 8. If we break 18 down as we did 3,095 in the place-value explanation, we get this:

$$
\begin{array}{lcl}
8 \times 1 & = & 8 \\
+ \quad 1 \times 10 & = & 10 \\
\hline
\text{Total} & & 18
\end{array}
$$

So, we see that there is 1 ten in 18. Therefore, in an addition problem, this 1 properly belongs in the second (tens) column.

In the same example, when we added the digits in the tens column (1 + 5 + 2 + 7 + 8), including the 1 carry digit, we got 23. *But these are all tens!* What we are actually adding is 10 + 50 + 20 + 70 + 80, which equals to 230. Two hundred-thirty breaks down like this:

$$
\begin{array}{rllr}
 & 0 \times 1 & = & 0 \\
+ & 3 \times 10 & = & 30 \\
 & 2 \times 100 & = & 200 \\
\hline
 & \text{Total} & & 230
\end{array}
$$

We see there are 2 hundreds in 230, so this 2 properly belongs in the third (hundreds) column.

If we are able to hold several numbers in our head at one time we can use our knowledge of place value to add several multi-digit numbers. Let's start with something relatively easy, adding 42 and 56.

What we do is *decompose* each number into is place-value parts; how many ones, how many tens, how many hundreds, etc.

Forty-two decomposes into 4 tens and 2 ones.

Fifty-six decomposes into 5 tens and 6 ones.

Adding the tens together, 4 + 5, we get 9 *tens*—or 90.

Adding the ones together, 2 + 6, we get 8 *ones*—or simply, 8.

To get the total we add the parts: 9 tens (90) with 8 ones (8): 90 + 8 = 98.

Of course, it gets more complicated with bigger numbers as well as with more numbers. You have to hold more numbers separately in your head, and that is more difficult.

Let's try another example. Add $48 + 137 + 64$.

Schematically, these numbers break down like this:

Number	Hundreds	Tens	Ones
48		4	8
137	1	3	7
64		6	4
Total	1	13	19

We start with the large parts of the numbers and work our way through the tens and finally to the ones. So in this case we note we have 1 hundred. We mentally file away 100.

Then we look at the tens and see we have 13. That means 1 hundred and 3 tens, 1 hundred plus 30. We add the 1 hundred to the other one and mentally file away that we now have 200 and 30 (230).

Finally we look at the ones and see we have 19. That's 1 ten and 9 ones. We add the 1 ten to 30 to get 40. Our grand total so far is $230 + 10 = 240$.

Then we tack on the 9 ones: $240 + 9 = 249$.

This method of adding by decomposition isn't just for doing addition mentally, you can also think of it as an alternative to the Standard Addition algorithm. Doing the above problem with pencil and paper, we would write out the schematic breakdown as above, and then interpret the Total row:

	Hundreds	Tens	Ones
Total	1	13	19

like this:

$$
\begin{array}{rcr}
19 \times 1 & = & 19 \\
+ \quad 13 \times 10 & = & 130 \\
1 \times 100 & = & 100 \\
\hline
\text{Total} & & 249
\end{array}
$$

I suspect that most people will be happy to stick with the Standard Addition algorithm that has served us so well for a long time, but if the decomposition alternative floats your boat, be my guest.

You-Try-It answers

1a. 312   1b. 267   1c. 1,681   1d. 1,475

Exercises 2.2

Add

1. 38 + 51

2. 86 + 57

3. 46 + 84

4. 119 + 47

5. 114, 53 and 72

6. 380 + 68 + 112 + 35

Add mentally by decomposition. Check your result.

7. 43 + 52

8. 37 + 49

9. 18 + 52 + 23

10. 53 + 118 + 34

2.3 Subtraction

Now that we've tackled addition we can go forward with subtraction. In addition the answer is called the *sum* or the *total*. When doing a subtraction problem, the answer is called the *difference*. There is more than a fair amount of vocabulary involved in subtraction. Much of it is unnecessary to know (namely, the names of the two numbers in a subtraction problem), but some of them you should be aware of. Here are several different ways I could describe the subtraction problem $10 - 2$:

- Ten subtract 2.
- Ten minus 2.
- Ten less 2.
- Two less than 10.

You might be surprised to hear that in other countries there are different "Standard algorithms" than the one we are taught using "borrowing." I like to amuse—or is it *confuse*?—my students by showing them the Standard Subtraction algorithm as I learned it... *in England*. If you're good I'll share it with you also, after explaining *our* Standard Subtraction algorithm.

We'll also see if decomposing the numbers will make subtraction easier, or come to the conclusion that sometimes it is best to "let sleeping dogs lie" ... as they say.

2.3.1 The Standard Subtraction algorithm

We know that subtraction is harder than addition, but it does have one advantage: we only subtract numbers in pairs. There's no limit to how many numbers we may be asked to add, but with subtraction, we're only doing one at a time—that is, subtracting *one* number from another.

And like with addition, the two numbers need to be arranged the same way: vertically right-aligned. For our first example, let's subtract numbers where there's no borrowing necessary. This should be easy.

Let's subtract 274 from 587. Note, this is written $587 - 274$. Unlike addition, where $2 + 3$ is the same as $3 + 2$, *order matters* in subtraction! We start by right-aligning (or "right-justifying") the numbers:

$$
\begin{array}{r}
5\ \ 8\ \ 7 \\
-\ \ 2\ \ 7\ \ 4 \\
\hline
\end{array}
$$

Next, we subtract the bottom digits from the ones above them, starting at the right-most, ones column and moving left one column at a time.

Four from 7 is 3. We write 3 in the ones place in the difference row:

$$
\begin{array}{r}
5\ \ 8\ \ 7 \\
-\ \ 2\ \ 7\ \ 4 \\
\hline
3
\end{array}
$$

Eight minus 7 is 1. Write 1 in the tens place in the difference row:

$$
\begin{array}{r}
5\ \ 8\ \ 7 \\
-\ \ 2\ \ 7\ \ 4 \\
\hline
1\ \ 3
\end{array}
$$

Five less 2 is 3. Write 3 in the hundreds place in the difference row:

$$
\begin{array}{r}
5\ \ 8\ \ 7 \\
-\ \ 2\ \ 7\ \ 4 \\
\hline
3\ \ 1\ \ 3
\end{array}
$$

(Notice how I'm sneaking in the different subtraction words to keep you on your toes.)

The answer is 313.

Big deal, you say. You know how to do this already; teach me how to do it when I'm forced to subtract a *larger* digit from a *smaller* one!

Okay, I yield to your demand. Let's subtract 87 from 635.

$$
\begin{array}{r}
6\ \ 3\ \ 5 \\
-\ \ \ \ \ 8\ \ 7 \\
\hline
\end{array}
$$

Our first job is to take 7 from 5. We can't do that. Seven is more than 5! *I get it.*

What we do is, once again, take advantage of our knowledge of place-value. Sure, we can't take 7 from 5, but that 5 is part of the number 635. There's a lot more to 635 than just the 5 ones. There's 3 tens, for example.

We could use one of those tens in order to pull off our ones subtraction of 7 from 5. If we borrow one of the 3 tens, we can add it to the 5 ones, giving us 15 ones. Now we can certainly subtract 7 ones from 15 ones: $15 - 7 = 8$. So, we're going to write 8 in the ones place in the difference row, the first digit in our answer, but we have to account for the fact that we borrowed 1 ten in order to do it. So we knock down the 3 tens we had by 1. We cross out 3 and replace it with a 2, above and a little to the left:

$$
\begin{array}{r}
2\ \ 1 \\
6\ \ \cancel{3}\ \ 5 \\
-\ \ \ \ 8\ \ 7 \\
\hline
8
\end{array}
$$

Next we move over to the tens column and try to perform the subtraction there: subtracting 8 from 2. Once again we run into this same problem: trying to take a bigger number from a smaller. Luckily, once again, we have some room to move because 635 also has 6 hundreds. If we borrow one of them and add it to the 2 tens remaining (after the first borrowing), we now have 12 tens.

(One hundred is 10 tens, so we're adding 10 *tens* to 2 *tens*—not *1 ten* to 2 tens—and this gives us the 12 *tens*. If this is confusing, you really don't need to worry about it. You have a 6 in hundreds place, of which you are going to borrow 1. In the tens place you have a 2. All you do is write the 1 before the 2, i.e., so you see the number 12 in the tens place; and subtract the 1 from the 6 in the hundreds column, leaving 5 hundreds.)

Now, you can subtract 8 from 12 to get 4, which you write in the tens place in the difference row... and of course, don't forget to decrease the 6 in the hundred's place by 1, replacing it with 5.

$$
\begin{array}{r}
5\ \ {}^{1}2\ \ 1 \\
\cancel{6}\ \ \cancel{3}\ \ 5 \\
-\ \ \ \ 8\ \ 7 \\
\hline
4\ \ 8
\end{array}
$$

You now have a 5 in the hundreds place from which you have nothing to subtract, so it gets carried down to the answer (difference) row. The answer is 548.

$$
\begin{array}{r}
5\ \ {}^{1}2\ \ 1 \\
\cancel{6}\ \ \cancel{3}\ \ 5 \\
-\ \ \ \ 8\ \ 7 \\
\hline
5\ \ 4\ \ 8
\end{array}
$$

Before we leave off and give you a chance to work some subtraction problems, let's tackle one more example: where you subtract from a number that has several consecutive zeros, such as 100, 400, or 2,000.

The principle is the same: borrowing from the place-value column to the left, but if you have to do this repeatedly, for instance if your subtracting 246 from 2,000, you want to first see how to do this in the simplest and quickest manner.

Example.   2000 – 246

Step 1.	Vertically right-align the numbers.	2 0 0 0 – 2 4 6
Step 2.	Clearly you are going to have to borrow. Further, clearly you are going to have to borrow more than just once for the ones column. You know you are going to need to borrow 1 ten for the ones column, but you don't have any tens—nor hundreds—to borrow from. *You are going to end up borrowing 1 thousand.* When all the borrowing is done, you will have 1 thousand, 9 hundreds, 9 tens and 10 ones. You put the 1 into the ones column to make 10, then replace the "200" with "199" in the thousands/hundreds/tens columns.	1 9 9 1 2 0 0 0 – 2 4 6

Step 3.	Do all the subtractions:
	
	    ones: $10 - 6 = 4$
	
	    tens: $9 - 4 = 5$
	
	    hundreds: $9 - 2 = 7$
	
	    thousands: $1 - 0 = 1$
	
	The answer is 1,754
		1 9 9 1
		2 0 0 0
		– 2 4 6
		———
		1 7 5 4

A lot is happening in *Step 2*. If you care to see the "plumbing," here it is:

1.   We borrow 1 ten from 0 tens which gives us 10 ones.
2.   Since we had no tens we first had to borrow 1 hundred—10 tens.
3.   Then we borrowed the 1 ten needed for the ones, which left us with 9 tens.
4.   But we had no hundreds to borrow from! So we had to borrow 1 thousand—10 hundreds—in order to get the 1 hundred we needed to borrow for the tens.
5.   After borrowing the 1 hundred from the 10 hundreds that we borrowed from the 2 thousands we had, we are left with 9 hundred, and 1 thousand.

This "cascading borrowing" is complicated but really not necessary to follow *in order to actually do the subtraction*. If you have successive zeros they are going to be replaced with 9s, and the first non-zero digit to the left of the successive zeros will be decreased by 1. **All you need do is follow the procedure as summarized in** *Step 2*, **i.e., replace 2000 with 199^{1}0.**

Here are some examples of how a number with successive zeros will be changed if you need to subtract from it:

$$2, \ 0 \ \ 0 \ \ 0 \ \rightarrow \ 1, \ 9 \ \ 9 \ \ {}^1 0$$

$$5, \ 0 \ \ 0 \ \ 7 \ \rightarrow \ 4, \ 9 \ \ 9 \ \ {}^1 7$$

$$1 \ 0, \ 0 \ \ 0 \ \ 0 \ \rightarrow \ 9, \ 9 \ \ 9 \ \ {}^1 0$$

Note, if the number you are subtracting ends in one or more zeros, the change necessary for the borrowing will differ from those shown here.

Now it's your turn.

You-Try-It 1

Subtract:

a)   Subtract 257 from 369

b)   364 minus 168

c) 1,037 – 852

d) 130,005 – 48,064

2.3.2 Subtraction by decomposition

Just as we did with addition, we are going to show a mental shortcut for subtraction. If this is more than you need or want to know, skip it. We are only going to do problems with numbers less than 100. You can really go nuts with this once you start working with larger and larger numbers. For now, I think it will be a significant achievement if you can do subtraction with numbers less than 100 using this mental decomposition. Feel free to experiment on your own by trying to do harder problems like, for instance, 327 – 184.

What we do is adjust (*round*) our numbers to the nearest ten, perform the subtraction of quantities of tens (such as 20, 70, 40, etc.), and then bring back the ones that we discarded by either adding them to or subtracting them from our preliminary estimate.

That sounds very complicated—and as I said feel free to ignore this section if you are happy with doing subtraction using the Standard Algorithm—but I hope when you see a problem done step-by-step you will agree that it really isn't that bad, and can be very useful if you can master it.

(If you are unfamiliar with *rounding* it is covered in the beginning of Chapter 4: Decimals.)

Example. Subtract 48 from 93

We reason: This problem would be a lot easier if it was just 90 – 50. I know *that*—it's 40!

Well, that's exactly how we start. We round the numbers to their nearest ten and do *that* subtraction. So, subtracting 50 from 90 to get 40 is the first step.

Preliminary answer:
$$90 - 50 = 40$$

Now, this "answer" will clearly need to be adjusted because ... *it's not the real problem!*

In the first place, we are subtracting from 93—not 90. Ninety-three is 3 more than 90. Our preliminary answer is too small by this 3; so we add it back in:

Working answer:
$$40 + 3 = 43$$

So our "working answer" is now 43. But, the number we are subtracting is actually 48, not 50.

Forty-eight is 2 less than 50. Our "working answer" is too small by this 2. We need to add it back in:

$$43 + 2 = 45$$

And this is the final answer. $93 - 48 = 45$.

Let's do one more:

Example. $67 - 28$

Step 1.	Round 67 to nearest ten: 70. Round 28 to nearest ten: 30. Do this subtraction to get a preliminary answer.	$70 - 30 = 40$
Step 2.	The number we are actually subtracting from is 67, which is 3 less than 70. Our preliminary answer is therefore too *big* by 3. We need to subtract this 3 from our preliminary answer.	$40 - 3 = 37$
Step 3.	The number we are actually subtracting is 28, not 30. It is 2 less than 30. Our working answer is therefore too *small* by 2. We need to add the 2 back into our working answer, giving us the final answer: 39.	$37 + 2 = 39$

You-Try-It 2

Subtract by mental decomposition.

a) Subtract 64 from 93

b) 87 minus 38

c) 67 – 32

d) 73 – 49

2.3.3 Alternate method for borrowing

Those wacky English, they do everything backward. They drive on the wrong side of the road. The first year of elementary school is Class 6 from which they proceed to Class 1. While the rest of the world uses the "right-hand rule" for describing the direction of current in the presence of a magnetic field, they use "Fleming's left-hand rule." They say "Happy Christmas" instead of "Merry Christmas."

Okay, you get the idea. And so it should come as no surprise to you that they have a backward way of borrowing in subtraction, too.

In English subtraction, instead of borrowing a ten from the digit to the left, decreasing it by 1, you *add* a 1 to the digit *beneath* it. We're taught to memorize this jingle: "Add a 10 to the top and a 1 to the bottom."

Here's an example comparing the two methods:

Example. Subtract: 83 – 37

Normal way		Backwards English way	
Step 1. Vertically right-align the numbers. We need to subtract 7 from 3 so we need to borrow a ten from the 8.	8 3 – 3 7 ———	*Step 1.* Same—but the borrowing will be done by adding a ten to the 3 tens in 37.	8 3 – 3 7 ———
Step 2. Write the 1 above and to the left of the 3, which we now consider 13. Subtract 7 from 13; write the result, 6, below in the answer row.	1 8 3 – 3 7 ——— 6	*Step 2.* (same)	1 8 3 – 3 7 ——— 6
Step 3. Decrease the 8 by 1; crossing it out and writing 7 above it and to the left. Subtract 3 from 7. Write the result, 4, below it in the answer row. The final result is 46.	7 1 8̶ 3 – 3 7 ——— 4 6	*Step 3.* Add 1 to the 3 below the 8. Cross out 3 and write 4. Subtract 4 from 8. Write the result, 4, below it in the answer row. The final result is 46.	1 8 3 – ⁴3̶ 7 ——— 4 6

You see, there is no difference between subtracting 7 – 3 and 8 – 4!

Now you can do math like an Englishman or Englishwoman. Try doing some of the exercises using this method, if you like. Practice driving on the wrong side of the road and you'll be all set. Happy days!

You-Try-It answers

1a. 112	1b. 196	1c. 185	1d. 81,941
2a. 29	2b. 49	2c. 35	2d. 24

Exercises 2.3

Subtract.

1. $87 - 13$ 2. $100 - 54$

3. $157 - 63$ 4. $95 - 48$

5. $2,083 - 974$ 6. $102,058 - 33,149$

7. $4,005 - 186$ 8. $10,054 - 9,255$

Subtract by decomposition.

9. $78 - 14$ 10. $95 - 67$

11. $100 - 33$ 12. $54 - 18$

2.4 Multiplication

Multiplication and division are more often than not done with a calculator. Nonetheless, everyone should know how to do both by hand in a pinch—after all, you never know when the power will go out and you're faced with a burning need to multiply 273 by 87 or divide 13,202 by 46. We will only show the Standard Algorithms for each. If you want a shortcut... use a calculator.

Multiplication vocabulary

The answer to a multiplication operation is called the *product*. Numbers that are multiplied to produce a product are called *factors*. 2 x 4 = 8. Two and four are *factors* of 8. The *product* of 2 and 4 is 8.

2.4.1 Times Table

There is much hand-wringing over the general poor state of mathematics readiness of students entering college. The over-reliance of the calculator in grade school is often given as one of the chief culprits; in particular, the use of the calculator to perform standard multiplications and divisions like multiplying 7 by 9, or dividing 96 by 12.

Eons ago, when I was a wee lad in Class 5 at the local Church school in London (translation: 2nd grade at the *public* school), there was no such thing as a calculator—indeed, the transistor had only recently been invented. The products of the single-digit numbers had to be memorized. We used printed Multiplication Tables to memorize them, reinforced by verbal and written drills.

The standard set of Multiplication Tables includes the products of all the numbers between 2 and 10, in every combination, of course.

But we were in England, don't forget, and so we had additional Times Tables to memorize: 11, 12 and believe it or not, 16. For your benefit, I am going to include 11, 12 and 16 in my Times Table. When multiplication is first taught in grade school there are separate tables for each number; one for the products of 2, one for those of 3, etc. I am combining all of them into one giant table. If you don't already know these products please do your best to learn at least the ones through 10.

To be able to do any math beyond basic math, such as algebra, you place yourself at a severe disadvantage if you do not *"know your Times Tables."* When you have thoroughly learned the Times Tables you have not just memorized the basic products, you also have basic quotients at your fingertips, such as what you get when 72 is divided by 6 (Division: next section).

Times Table

	2	3	4	5	6	7	8	9	10	11	12	16
2	4	6	8	10	12	14	16	18	20	22	24	32
3	6	9	12	15	18	21	24	27	30	33	36	48
4	8	12	16	20	24	28	32	36	40	44	48	64
5	10	15	20	25	30	35	40	45	50	55	60	80
6	12	18	24	30	36	42	48	54	60	66	72	96
7	14	21	28	35	42	49	56	63	70	77	84	112
8	16	24	32	40	48	56	64	72	80	88	96	128
9	18	27	36	45	54	63	72	81	90	99	108	144
10	20	30	40	50	60	70	80	90	100	110	120	160
11	22	33	44	55	66	77	88	99	110	121	132	176
12	24	36	48	60	72	84	96	108	120	132	144	192
16	32	48	64	80	96	112	128	144	160	176	192	256

If you were not taught the Times Tables in school, I strongly suggest that you make up your own drills and invest the time to learn them now.

2.4.2 The Standard Multiplication algorithm

Now let's go straight to the step-by-step, how-to, format to explain how to do multiplication of numbers with more than one digit:

Example. Multiply 8 x 52

Step 1.	Vertically right-align the numbers. It is easier if you write the number with the fewer digits on the bottom.	5 2 × 8 ⎯⎯⎯⎯⎯
Step 2.	We start by multiplying the digit in the ones place of the upper factor, 2, by 8. 8 x 2 = 16. 16 is 10 + 6. 1 ten and 6 ones. We write 6 in the ones column in the product row, and "carry" the 1 ten into the tens row, placing it above and to the left of the digit occupying the tens place, the 5.	¹5 2 × 8 ⎯⎯⎯⎯⎯ 6
Step 3.	We are now working with *tens*! Multiply the 5 by the 8. 8 x 5 = 40. Add the 1 carry *(ten)*: 40 *(tens)* + 1 *(ten)* = 41 *(tens)*.	¹5 2 × 8 ⎯⎯⎯⎯⎯ 6
Step 4.	Since these were the last digits to multiply, simply write 41 into the product row to the left of the 6. The answer is 416.	¹5 2 × 8 ⎯⎯⎯⎯⎯ 4 1 6

When we have two or more digits in the smaller of the factors, the process becomes more complicated. If you are multiplying by a 2-digit number, you need three product rows: two partial product rows that you then *add* to produce the final product in the last row.

If the smallest factor has 3 digits, you will have 3 partial product rows and a 4th that will be the final product produced by adding the three partial product rows above it.

Example. Multiply 47 x 357

Step 1.	Right-align the numbers. It is easier if you write the number with the fewer digits on the bottom.	3 5 7 × 4 7
Step 2.	We start by multiplying by the digit in the ones place of the bottom factor: 7. First multiply it by the digit in the ones place of 357: 7. 7 x 7 = 49. 49 is 40 + 9. 4 tens and 9 ones. We write 9 in the ones column in the product row, and carry the 4 tens into the tens row, placing it above and to the left of the digit occupying the tens place: the 5.	3 ⁴5 7 × 4 7 ———— 9
Step 3.	We now multiply the next digit in the upper factor, 5 (the digit in the tens place of 357), by the 7 from the bottom factor. 7 x 5 = 35. Add the 4 carry: 35 + 4 = 39. We write the 9 in the tens place in the product row, and carry the 3 to the hundreds place in the upper factor, 357, placing it above and to the left of the digit occupying that place, 3.	³3 ⁴5 7 × 4 7 ———— 9 9

Step 4.	We now multiply the next and last digit in the upper factor, 3 (the digit in the hundreds place of 357), by the 7 from the bottom factor.	^{3}3 ^{4}5 7
	7 x 3 = 21.	$\times$ 4 7
	Add the 3 carry: 21 + 3 = 24.	———————
	Since this was the last digit to multiply in the upper factor, we just write 24 to the left of the 99 in the first partial product row.	2 4 9 9

| *Step 5.* | We now multiply the upper factor, 357, by the 2nd digit of the lower factor (47): 4. We will write these products in a second partial product row, below the first one we've just completed. Also, we will start filling in the digits in the tens place, not the ones, because we are actually multiplying by 40, not 4. We can write a 0 in the ones place to remind us, if we like. | ^{3}3 ^{4}5 7
 $\times$ 4 7
 ———————
 2 4 9 9
           0 |

| *Step 6.* | Multiply the first (ones) digit of 357 by 4: 4 x 7 = 28. Write the 8 in the tens place in the new partial product row. Carry the 2 by writing it above and to the left of the 5. (Ignore the carry digits already written there by the first row of multiplications. Cross them out or erase them.) | 3 ^{2}5 7
 $\times$ 4 7
 ———————
 2 4 9 9
         8 0 |

| *Step 7.* | Multiply the second (tens) digit of 357 by 4: 4 x 5 = 20. Add the carry digit, 2. 20 + 2 = 22. Write the 2 in the hundreds place in the new partial product row. Carry the other 2 by writing it above and to the left of the 3, the digit in the hundreds place of the upper factor. | ^{2}3 ^{2}5 7
 $\times$ 4 7
 ———————
 2 4 9 9
     2 8 0 |

| Step 8. | Multiply the third (hundreds) digit of 357 by 4: 4 x 3 = 12. Add the carry digit, 2. 12 + 2 = 14. Since there are no more digits in the upper factor, write 14 in the partial product row, to the left of the 280. | ^{2}3 ^{2}5 7
×   4 7
2 4 9 9
1 4 2 8 0 |
| Step 9. | We get the final answer by adding the 2 partial product rows.

2,499 + 14,280 = 16,779. | ^{2}3 ^{2}5 7
×   4 7
2 4 9 9
1 4 2 8 0
1 6 7 7 9 |

You-Try-It 1

a) Multiply 7 by 83

b) 14 × 38

c) 36×74 d) 43×216

You-Try-It answers

1a. 581   1b. 532   1c. 2,664   1d. 9,288

Exercises 2.4

Multiply.

1. 34×7

2. 58×6

3. 253×8

4. 43×16

5. 37×24

6. 243×28

2.5 Division

Division is considered the most difficult of the four arithmetic operations. Like multiplication of large numbers, you will not be frowned upon for going to the calculator when you need to do division. Division, or *long division* as it's called, can be a tedious operation, and it is easy to slip up if you're not very careful.

In a multiplication problem we basically are either multiplying or adding numbers. In long division, we end up dividing, multiplying and subtracting to get the final result. And we know how mindful we have to be with subtracting when there's borrowing involved... which there usually is.

Division vocabulary

Let's start with this: 4 x 3 = 12. Multiplication. We should be comfortable with this by now. The numbers 3 and 4 are *factors* of 12. We can take the same three numbers and turn this into a statement of division:

$$12 \div 3 = 4$$

The "$\div$" symbol represents division. We first point out that $12 \div 3$ is not the same as $3 \div 12$. *Order matters in division*, just as it does in subtraction. For this reason, we have to name the numbers in a division problem.

$$12 \quad \div \quad 3 \quad = \quad 4$$

Dividend   Divisor   Quotient

The *dividend* divided by the *divisor* equals the *quotient*.

2.5.1 Long division

Long division is what we do when we need to divide one number by another and the numbers involved are not those memorized by the multiplication tables. For instance, if you were asked to divide 12 by 3, you *should* know the answer is 4 because you *should* know your 3- and 4-times multiplication tables. You *should* know that 3 times 4 is 12, and so there is no need to do long division (or go to the calculator).

However, if you needed to find 136 ÷ 8, that's a different story. One hundred thirty-six does not appear in the conventional 8-times table, which goes up to 10 or 12 times 8 (or 16, if you were lucky enough to be raised in England). So, it is expected that you will do long division or just resort to the calculator. Of course, where I used the word *should* several times in the paragraph above, you and I both know the truth: Many people do not know their times tables. I could write an entire book about the effects on education of the calculator... but that's not the purpose of this one. If you don't know your times tables, you already know that you're starting with a handicap. Don't worry too much if you have to resort to using the calculator to do the elementary divisions and multiplications involved in doing long division. I forgive you.

Long division is a 4-step algorithm, meaning, the same 4 steps are repeated several times until we reach the final answer, the quotient. We'll start with a division problem with a 1-digit divisor, then do one with a *remainder*, and then do one with a 2-digit divisor.

Example. Divide 504 by 6

Step 1.	Align the numbers as shown. The divisor is outside the "box". The dividend is inside.	$6\overline{)5\ 0\ 4}$
Step 2.	We start by examining the dividend 504, one digit at a time starting from the left with 5. We ask the question: Is 5 greater than the divisor (6)? The answer is no. We then proceed to the second digit in the divisor, 0, and ask the question, is 50 greater than the divisor? The answer is Yes. We then divide 6 into 50. We ask what is the greatest number of 6s in 50? The answer is 8, because 8 x 6 = 48, less than 50 (whereas 9 x 6 = 54, is greater than 50, so 9 is too big). We write 8 on top as the first digit in the quotient. We write it over the 0, because we divided 6 into 50. This completes the first step in the algorithm.	$\begin{array}{r} 8 \\ 6\overline{)5\ 0\ 4} \end{array}$

Step 3.	We now multiply the 8 in the quotient by the divisor, 6, and write the result, 48, under the 50. This completes the 2nd step in the algorithm.	$$\begin{array}{r} 8 \\ 6\,\overline{\smash{)}5\ \ 0\ \ 4} \\ 4\ \ 8 \end{array}$$
Step 4.	The 3rd step in the algorithm is to subtract the 48 from the 50 above it, and write the difference just as we would if this were the subtraction problem $50 - 48$.	$$\begin{array}{r} 8 \\ 6\,\overline{\smash{)}5\ \ 0\ \ 4} \\ \underline{4\ \ 8} \\ 2 \end{array}$$
Step 5.	The 4th and final step in the algorithm is to bring down the next digit in the dividend, 4, and put it next to the 2.	$$\begin{array}{r} 8 \\ 6\,\overline{\smash{)}5\ \ 0\ \ 4} \\ \underline{4\ \ 8} \\ 2\ \ 4 \end{array}$$
Step 6.	We now begin the 4-step algorithm again. We divide our divisor 6, into 24. Six goes into 24 four times. We write 4 on top in the quotient.	$$\begin{array}{r} 8\ \ 4 \\ 6\,\overline{\smash{)}5\ \ 0\ \ 4} \\ \underline{4\ \ 8} \\ 2\ \ 4 \end{array}$$
Step 7.	The 2nd step in the algorithm is to multiply the 4 in the quotient by our divisor, 6, and write it below the 24.	$$\begin{array}{r} 8\ \ 4 \\ 6\,\overline{\smash{)}5\ \ 0\ \ 4} \\ \underline{4\ \ 8} \\ 2\ \ 4 \\ 2\ \ 4 \end{array}$$
Step 8.	The 3rd step in the algorithm is to subtract. We see we are just subtracting 24 from itself, which leaves us with nothing. Since there are no more digits in the dividend, we are done. $504 \div 6 = 84$. The quotient is 84.	$$\begin{array}{r} 8\ \ 4 \\ 6\,\overline{\smash{)}5\ \ 0\ \ 4} \\ \underline{4\ \ 8} \\ 2\ \ 4 \\ \underline{2\ \ 4} \\ 0 \end{array}$$

Note: $504 \div 6 = 84$ also means that $504 \div 84 = 6$, and also that $6 \times 84 = 504$.

We've found that 6 *goes into* 504 eighty-four times. Six divides *evenly* into 504, because there is *no remainder* when we divide 504 by 6. Six is therefore a factor of 504. And 84 is also a factor of 504, because *it* divides evenly into 504, going into it 6 times, with no remainder.

You-Try-It **1**

Divide 224 by 7.

2.5.2 Division with remainders

Now let's tackle long division when the divisor is *not* a factor of the dividend, i.e., when it does not divide evenly, when there is a *remainder*.

Example. $387 \div 8$

Step 1.	Align the numbers as shown. The divisor is outside the "box". The dividend is inside.	$8\,)\overline{\,3\ \ 8\ \ 7\,}$

Step 2.	We need to go to the 2nd digit in the dividend to find a number greater than our divisor. Thirty-eight is greater than 8. Eight goes into 38 at most 4 times. Write 4 as the first digit in the quotient, above the 8 of the dividend.	$\begin{array}{r} 4 \\ 8\overline{)3\ 8\ 7} \end{array}$
Step 3.	We now multiply the 4 in the quotient by the divisor, 8, and write the result, 32, under the 38.	$\begin{array}{r} 4 \\ 8\overline{)3\ 8\ 7} \\ 3\ 2 \end{array}$
Step 4.	Subtract the 32 from the 38 above it, and write the difference, 6, underneath.	$\begin{array}{r} 4 \\ 8\overline{)3\ 8\ 7} \\ 3\ 2 \\ \hline 6 \end{array}$
Step 5.	Bring down the next, and last, digit in the dividend, 7, and put it next to the 6.	$\begin{array}{r} 4 \\ 8\overline{)3\ 8\ 7} \\ 3\ 2 \\ \hline 6\ 7 \end{array}$
Step 6.	How many times does 8, the divisor, go into 67? It goes 8 times. Write 8 in the quotient next to the 4 already there.	$\begin{array}{r} 4\ 8 \\ 8\overline{)3\ 8\ 7} \\ 3\ 2 \\ \hline 6\ 7 \end{array}$
Step 7.	Multiply the 8 in the quotient by our divisor, 8, to get 64, and write it below the 67.	$\begin{array}{r} 4\ 8 \\ 8\overline{)3\ 8\ 7} \\ 3\ 2 \\ \hline 6\ 7 \\ 6\ 4 \end{array}$

Step 8.	Subtract 64 from 67. The result is 3. There are no more digits in the dividend to bring down. The division is done. There is a remainder of 3. Write "r3" next to the quotient. The final result of 387 divided by 8 is 48 r3.	$\begin{array}{r} 4\ \ 8\ \ \text{r3} \\ 8\,\overline{\smash{)}\,3\ \ 8\ \ 7} \\ 3\ \ 2 \\ \hline 6\ \ 7 \\ 6\ \ 4 \\ \hline 3 \end{array}$

(An alternate way to express a quotient with a remainder is to use fractions, which we will learn in the next chapter. In the above example, if were using fractions we would write the quotient as the *mixed number,* $48\frac{3}{8}$.)

You-Try-It 2

a) $185 \div 7$ 　　　　　　　　　　b) $789 \div 5$

2.5.3 Division with larger divisors

The good news is that we now know everything about long division. There are no more rules to learn. The only thing left to do is practice with bigger numbers—specifically with divisors having more than 1 digit. We will also use this problem to show how to check the answer to a division problem.

Example. Divide 811 by 23

Step 1.	Align the numbers in the long division box as normal: divisor outside, dividend inside.	23) 8 1 1
Step 2.	We need to go to the 2nd digit in the dividend to find a number greater than our divisor, 23. How many times does 23 go into 81? We see it can't be 4, because 4 x 20 = 80 and 4 x 3 = 12; so 4 x 23 = 80+12 is clearly greater than 81. We thus reason that it should go in 3 times. (We may resort to using a calculator to do these tests.) Write 3 above the 1 in 81 in the quotient.	3 23) 8 1 1
Step 3.	We now multiply the 3 in the quotient by the divisor, 23, and write the result, 69, under the 81. (Note, if we had guessed wrong above and written 4 as the first digit in the quotient, we would catch our error in *this* step when we multiplied 4 by 23 and realized the result, 92, was greater than 81.)	3 23) 8 1 1 6 9
Step 4.	Subtract the 69 from the 81 above it, and write the difference, 12, underneath.	3 23) 8 1 1 6 9 — 1 2
Step 5.	Bring down the next, and last, digit in the dividend, 1, and put it next to the 12.	3 23) 8 1 1 6 9 — 1 2 1

Step 6.	How many times does 23, the divisor, go into 121? This might take some trial-and-error multiplication work (assuming you're not using a calculator). It goes 5 times. Write 5 in the quotient next to the 3 already there.	$$\begin{array}{r} 3\ 5 \\ 2\ 3\ \overline{\smash{)}8\ 1\ 1} \\ 6\ 9 \\ \hline 1\ 2\ 1 \end{array}$$
Step 7.	Multiply the 5 in the quotient by our divisor, 23, to get 115, and write it below the 121.	$$\begin{array}{r} 3\ 5 \\ 2\ 3\ \overline{\smash{)}8\ 1\ 1} \\ 6\ 9 \\ \hline 1\ 2\ 1 \\ 1\ 1\ 5 \end{array}$$
Step 8.	Subtract 115 from 121. The result is 6. There are no more digits in the dividend to bring down. The remainder is 6. Write "r6" next to the quotient. The final result of 811 divided by 23 is 35 r6.	$$\begin{array}{r} 3\ 5\ \text{r6} \\ 2\ 3\ \overline{\smash{)}8\ 1\ 1} \\ 6\ 9 \\ \hline 1\ 2\ 1 \\ 1\ 1\ 5 \\ \hline 6 \end{array}$$

2.5.4 Checking the answer to a division problem

Recall earlier when we showed that any division problem can be turned into a multiplication problem? The division problem

$$\underset{\text{Dividend}}{12} \quad \div \quad \underset{\text{Divisor}}{3} \quad = \quad \underset{\text{Quotient}}{4}$$

can be written as

$$\underset{\text{Divisor}}{3} \quad \times \quad \underset{\text{Quotient}}{4} \quad = \quad \underset{\text{Dividend}}{12}$$

So, we see that Divisor x Quotient (must) = Dividend. And indeed, this is how we check our answer after doing a division problem.

But what about if there is a remainder?   What if the division problem was 14 ÷ 3, rather than 12 ÷ 3?

Let's first do the division problem...

```
            4   r2
       3 | 1   4
           1   2
          ———————
               2
```

So, 14 ÷ 3 = 4 r2.

Here's how we adjust the check:

12 ÷ 3 = 4		
3 × 4 = 12		
Divisor Quotient Dividend		

$\longrightarrow$

14 ÷ 3 = 4 r2			
3 × 4 + 2 = 14			
Divisor Quotient Remainder Dividend			

So, in general, the scheme for checking a division problem with a remainder is:

Divisor x Quotient + Remainder = Dividend

Let's show this works by checking the answer to our previous division example:

$$811 \div 23 = 35\,r6$$

$$23 \quad \times \quad 35 \quad + \quad 6 \quad \overset{?}{=} \quad 811$$

Divisor   Quotient   Remainder   Dividend

Step 1.	Multiply divisor, 23, by quotient, 35.	$23 \times 35 = 805$
Step 2.	Add the remainder, 6. The result is equal to the dividend, 811. Our check is complete.	$805 + 6 = 811$

You-Try-It 3

Divide **and check**:

a) $30 \div 4$

b) $228 \div 12$

c) $442 \div 26$

d) $596 \div 32$

You-Try-It answers

1. 32

2a. $26\frac{3}{7}$

2b. $157\frac{4}{5}$

3a. $7\frac{1}{2}$

3b. 19

3c. 17

3d. $18\frac{5}{8}$

Exercises 2.5

Divide and check.

1. $324 \div 6$

2. $273 \div 7$

3. $338 \div 4$

4. $250 \div 9$

5. $455 \div 13$

6. $544 \div 17$

7. $473 \div 43$

8. $527 \div 31$

9. $305 \div 12$

10. $481 \div 26$

Chapter 3 Fractions

3.1 Basic concepts and properties

Generally speaking, *fraction* is the word we use when talking about a part of a whole, though its meaning can be extended to include quantities of more than 1 whole.

Consider a pizza pie cut into eight slices. One piece (slice of the pie) is one *eighth* of the whole pie, which written as a fraction looks like this:

$$\frac{1}{8}$$

Two pieces would represent two-eighths of the pie, written as a fraction like this: $\frac{2}{8}$.

Three slices would represent three-eighths of the pie, written $\frac{3}{8}$. And so on.

The number on the bottom of the fraction bar is called the *Denominator*. The denominator tells us how many pieces (all of equal size) the whole is split into.

The number on top is called the *Numerator*. The numerator describes how many pieces of the whole are counted.

A pizza pie could be divided into 2, 4, or 6 pieces as easily as 8 pieces. If the pie were cut into 6 (equal) slices, a denominator of 6 would be used to represent one or more of the slices. To represent 1 slice of a pie cut into 6 pieces as a fraction, the numerator would be 1 and the denominator 6; written: $\frac{1}{6}$. Five slices would then be $\frac{5}{6}$.

These examples of how slices of a pie are considered as part of a whole illustrate the central idea behind fractions: *Parts of a whole*. However, in math we seldom allow a good idea to stand still. The idea of a fraction is extended to include parts of more than one whole.

3.1.1 Proper and improper fractions

Consider a gathering where one pizza pie is not enough. Let's say you need three pies to accommodate all the people. Assuming the pies are cut into eighths, how do you represent, say, 9 slices?

There are two ways this could be done. We could continue as before, consider the denominator to represent how many (equal) pieces there are in one pie, which would be 8—or, perhaps we should consider the denominator to be the total number of slices in the three pies: 24?

Now, there is no ambiguity about the value of the numerator: the only number that makes sense to use is 9, for the 9 slices. But it seems like a judgment call is in order to decide whether we should use 8 or 24 for the denominator.

Actually, it really isn't much of dilemma. The denominator represents how many pieces **one** whole thing is split into. In this instance, that *one whole thing* is **one** pie, and 1 pie is divided into 8 slices, so the denominator remains 8, regardless of how many pies there are.

Therefore, the fraction we use to represent 9 slices out of the 3 pies cut into eighths is

$$\frac{9}{8}$$

Note, the same fraction $\frac{9}{8}$ would be used whether there were two, three, or more than three pies (as long as each pie is divided into eight slices).

Any fraction where the numerator is greater than the denominator has a value that is greater than 1, and we call such a fraction an *improper fraction*.

***Definition.* Improper fraction:** A fraction where the numerator is greater than the denominator. An improper fraction has a numerical value that is greater than 1.

[Note however, it is true that if we really want to consider how much of the *available* pizza is represented by 9 slices, that quantity is the *ratio* of 9 slices to 24 slices, which can indeed be written as the fraction $\frac{9}{24}$. Ratios are often written as fractions, but there are other ways to represent them.]

'Normal' fractions, where the numerator is less than the denominator, are known as ***proper fractions***.

3.1.2 Simplifying fractions

Before we learn how to add, subtract, multiply and divide fractions, we have to learn some basic things about working with them. Taken together, these various manipulations we perform on them are all called *simplifications*.

Simplifying is a catch-all phrase that is used throughout mathematics and means different things depending on the kind of math we are doing. In almost every case, however, when a mathematical expression of any kind is *simplified*, it is rewritten in a form that is more compact. This usually means that fewer strokes of a pencil are required to write it, or, in the case of fractions, that smaller numbers are used to write it.

If a fraction were one number, it *couldn't* be written using a different number. After all, a different number is ... *a different number!*—and so *must* represent a different quantity.

But a fraction is composed of two numbers, and it turns out that often (more often than not, in fact) a fraction can be written using smaller numbers—and in fact, in some cases can be written in simplest form as one (whole) number.

3.1.2.1 Equivalent fractions

Consider again our pizza pie divided into 8 slices.

If you eat four of the slices, you have eaten half of the pie. Guess what? *Half* is a fraction! If you eat 4 slices out of 8 it is exactly the same as having eaten one-half of the pie. The fraction *one-half* is written $\frac{1}{2}$, so therefore 4 slices out of 8, $\left(\frac{4}{8}\right)$, must equal one-half, $\left(\frac{1}{2}\right)$!

Or,

$$\frac{4}{8} = \frac{1}{2}$$

Note, it is true that when we write $\frac{4}{8}$ as its fractional equivalent $\frac{1}{2}$, we have lost something: the information regarding how the whole was originally split, i.e., into 8 slices. But in math we seldom care about that, because the more fundamental relation is intact: *the amount of the whole represented by the part*. What matters most is *this* quantity. Think about it: does it really matter how you count the amount of pizza you ate? The quantity is the same regardless of whether you consider it 4 slices or half a pie. Your stomach really doesn't care.

This type of *simplifying*, writing $\frac{4}{8}$ as its equivalent $\frac{1}{2}$, is called *reducing* the fraction. It's more formal name is *reducing to lowest terms*, but we typically just use the one word: *reduce*.

The fractions $\frac{4}{8}$ and $\frac{1}{2}$ are called *equivalent* fractions because they represent exactly the same quantity. If a pie were cut into 10 slices, it should be clear that the fraction $\frac{5}{10}$ is also equivalent to $\frac{4}{8}$ and to $\frac{1}{2}$. Is there any limit to how many different ways a fraction can be written—or, put another way, is there a limit to how many equivalent fractions exist for any given fraction?

The answer is No. There are an infinite number of fractions equivalent to any given fraction, such as $\frac{1}{2}$.

You may be wondering, how can we create equivalent fractions, especially for ones that aren't as simple as $\frac{1}{2}$, such as, say, $\frac{9}{24}$?

You might think that if you just add the same number to both of the numbers in a fraction, this might work.

Let's try it:

If we add 1 to both numerator and denominator of $\frac{1}{2}$ we get the fraction $\frac{2}{3}$.

$$\frac{1+1}{2+1} = \frac{2}{3}$$

Instinctively we know this is wrong. $\frac{2}{3}$ is the fraction two-thirds. It is not the same as $\frac{1}{2}$.

In terms of our pizza pie, this would be dividing the pie into three *equal* slices (not that easy to do, if you've ever tried it!), and eating two of them. Instinctively we know this is more than cutting the pie into two halves and eating one of them.

So, adding the same number to numerator and denominator changes the value of the fraction, therefore it does not create an equivalent fraction. However, if we *multiply* both numbers in a fraction by the same number, this does work. In fact, this is how we create equivalent fractions: we multiply the numerator and denominator by the same number—any number at all, no matter how big or small.

When the numerator and denominator of a fraction are multiplied by the same number, the resulting fraction is an *Equivalent fraction*.

Example: Creating equivalent fractions

$$\frac{1 \times 4}{2 \times 4} = \frac{4}{8} \qquad \frac{1 \times 3}{2 \times 3} = \frac{3}{6} \qquad \frac{1 \times 5}{2 \times 5} = \frac{5}{10} \qquad \frac{9 \times 5}{8 \times 5} = \frac{45}{40}$$

You-Try-It 1

Create three equivalent fractions for the following:

a) $\frac{3}{5}$

b) $\frac{2}{3}$

3.1.2.2 Reducing fractions

Now that we have an understanding of *equivalent fractions*, we are prepared to both appreciate the value of re-writing a fraction with the *smallest* numbers and to learn how to do this, i.e, *how to reduce a fraction.*

After all, would you really want to work with the fraction $\frac{651}{1519}$ if you knew that it has the exact same value as $\frac{3}{7}$?

Reducing a fraction to lowest terms means re-writing it as the equivalent fraction with the smallest numbers.

Recall that to undo multiplication we divide. To reduce a fraction we divide the numerator and denominator by the same number. The key is to find that one number that divides evenly into both the numerator and denominator, *and*, to find the largest such number so that the resulting new numerator and denominator will be the smallest possible for that particular fraction. This number, by the way, is known as the **Greatest Common Factor** (GCF) of the numbers.

For example, let's examine the fraction $\frac{36}{48}$.

Note that there are several equivalent fractions to $\frac{36}{48}$ that have smaller numbers, but when we say *reduce* $\frac{36}{48}$, we implicitly mean reduce it to that equivalent fraction that has the smallest numbers—i.e., to its *lowest terms*.

Noting that 36 and 48 are both even numbers, they are both divisible by 2. If we start there, we get this result:

$$\frac{36 \div 2}{48 \div 2} = \frac{18}{24}$$

Now, we have, technically speaking, *reduced* $\frac{36}{48}$, but we know that $\frac{18}{24}$ cannot be the smallest equivalent fraction to $\frac{36}{48}$ because, for one thing, both 18 and 24 are still even numbers, and so can be divided by 2. So, we have not reduced $\frac{36}{48}$ to its *lowest terms*, which as said before, is what we really mean when we say *reduce*.

It turns out that this particular fraction, $\frac{36}{48}$, reduces to $\frac{3}{4}$. Different people will arrive at this in different ways. Some of us may just continue to divide by 2 until we hit $\frac{9}{12}$. Then, recognizing that 9 is an odd number and so not divisible by 2, we might try dividing by the next whole number, 3:

$$\frac{9 \div 3}{12 \div 3} = \frac{3}{4}$$

Others, proficient in the times tables up to 12, may recognize right of the bat that both 36 and 48 are divisible by 12, and in fact 12 is the GCF of 36 and 48.

Those people will reduce $\frac{36}{48}$ in one clean step:

$$\frac{36 \div 12}{48 \div 12} = \frac{3}{4}$$

Thus revealed: the advantages to knowing one's times tables! Of course, some of you will need to use the calculator to test divisibility in order to reduce a fraction. The key lesson here is not so much about which method you use to reduce a fraction (which depends on your particular set of math skills), but rather *what it is you have to do*.

In essence, to reduce a fraction you have to find the largest number that divides evenly into both numerator and denominator (the GCF).

Example. Reduce the fraction

a) $\frac{9}{15}$

b) $\frac{6}{30}$

The GCF of 9 and 15 is 3:

$$\frac{9 \div 3}{15 \div 3} = \frac{3}{5}$$

$\frac{9}{15}$ reduces to $\frac{3}{5}$

The GCF of 6 and 30 is 6:

$$\frac{6 \div 6}{30 \div 6} = \frac{1}{5}$$

$\frac{6}{30}$ reduces to $\frac{1}{5}$

c) $\dfrac{350}{500}$

d) $\dfrac{24}{9}$

The GCF of 350 and 500 is 50:

$$\dfrac{350 \div 50}{500 \div 50} = \dfrac{7}{10}$$

The GCF of 24 and 9 is 3:

$$\dfrac{24 \div 3}{9 \div 3} = \dfrac{8}{3}$$

$\dfrac{350}{500}$ reduces to $\dfrac{7}{10}$

$\dfrac{24}{9}$ reduces to $\dfrac{8}{3}$

Let's take a closer look at examples (c) and (d) above.

In (c), $\dfrac{350}{500}$, both numerator and denominator end with a zero. This means they are divisible by 10. We refer to these as *trailing zeros*. When reducing a fraction, equal numbers of trailing zeros in numerator and denominator can be ignored, i.e, *crossed-out*, as a first step in the reducing process.

In our example the numerator has one trailing zero and the denominator two. Therefore you can cross out *only the one trailing zero that is common to both*. The reducing could be worked like this:

$$\dfrac{350}{500} \rightarrow \dfrac{35\cancel{0}}{50\cancel{0}} \rightarrow \dfrac{35}{50} \rightarrow \dfrac{35 \div 5}{50 \div 5} \rightarrow \dfrac{7}{10}$$

To reduce the fraction $\dfrac{6200}{15000}$ you would be wise to cross out two trailing zeros in each and start your reducing with $\dfrac{62}{150}$:

$$\dfrac{6200}{15000} \rightarrow \dfrac{62\cancel{00}}{150\cancel{00}} \rightarrow \dfrac{62}{150} \rightarrow \dfrac{62 \div 2}{150 \div 2} \rightarrow \dfrac{31}{75}$$

Example (d) , $\dfrac{24}{9}$ is an improper fraction. The first thing we want to point out about this fraction is that there is no difference in how we reduce when the numerator is greater than the denominator. We know that the quantity represented is greater than 1, but that has no effect on how we reduce.

$\dfrac{24}{9}$ reduces to $\dfrac{8}{3}$, and $\dfrac{8}{3}$ is greater than 1. Big deal.

3.1.3 Converting between mixed numbers and improper fractions

Sometimes we may want to write an improper fraction as what we call a *Mixed Number*.

Consider two pizza pies each cut into eighths. There is a total of 16 slices. The number of slices in one pie and half of the other one is 8 + 4 = 12 slices. As an improper fraction this is $\frac{12}{8}$.

Since both 12 and 8 are divisible by 4, we easily see that $\frac{12}{8}$ reduces to $\frac{3}{2}$.

So, the improper fraction $\frac{3}{2}$ is the same as one-and-a-half, or $1\frac{1}{2}$. This is a *mixed number*, it contains a whole number with a fraction next to it.

The mixed number $13\frac{2}{5}$ means 13 plus $\frac{2}{5}$. So, the addition expression $13 + \frac{2}{5}$ is identical to $13\frac{2}{5}$.

1. Converting a mixed number to an improper fraction

Example. Convert $13\frac{2}{5}$ to an improper fraction

Step 1.	The denominator of the improper fraction will be the same as the denominator of the fraction in the mixed number.	$13\frac{2}{5} = \frac{?}{5}$
Step 2.	Multiply the denominator by the whole number.	$5 \times 13 = 65$
Step 3.	Add the numerator.	$65 + 2 = 67$
Step 4.	This sum is the numerator of the improper fraction.	$13\frac{2}{5} = \frac{67}{5}$

You-Try-It 2

Change to an improper fraction:

a) $5\frac{2}{3}$

b) $11\frac{3}{7}$

2. Converting an improper fraction to a mixed number

Example. Convert $\frac{67}{5}$ to a mixed number

Step 1.	The denominator of the fraction in the mixed number will be the same as the denominator of the improper fraction.	$\frac{67}{5} = \Box\frac{?}{5}$
Step 2.	Divide the numerator by the denominator, using standard quotient and remainder scheme.	$67 \div 5 = 13\,r2$
Step 3.	The quotient is the whole-number part of the mixed number.	$\frac{67}{5} = 13\frac{?}{5}$
Step 4.	The remainder is the numerator of the fraction in the mixed number.	$\frac{67}{5} = 13\frac{2}{5}$

You-Try-It 3

Change to a mixed number:

a) $\frac{37}{4}$

b) $\frac{99}{6}$

It turns out that once we get to Algebra in mathematics, we seldom, if ever, write improper fractions as mixed numbers. Whether you use an improper fraction or a mixed number depends on the context of the math problem. In real life, using a mixed number will probably make more sense. ("You ate one-and-a-half pies!" sounds more natural than, "You ate three-halves of pie!")

3.1.4 Reciprocals

Looking back at our reducing examples in 3.1.2.2, example (d) was the improper fraction $\frac{24}{9}$. Did it remind you of $\frac{9}{24}$, the fraction we said represented the ratio of 9 slices to the total in the three pies? There is a relationship between these two fractions. The numerators and denominators are switched, or flipped. The numerator of $\frac{24}{9}$ is the same as the denominator of $\frac{9}{24}$, and the denominator of $\frac{24}{9}$ is the same as the numerator of $\frac{9}{24}$.

These two fractions are called *reciprocals*.

Definition. **Reciprocal:** The reciprocal of any fraction $\frac{a}{b}$ is the fraction $\frac{b}{a}$ [2]

What about the reciprocal of a whole number, such as 2, 3, or 798?

Before we answer, we need to explain a little more about *simplifying*.

So far the only simplifying we have talked about is *reducing*, but another type of simplifying permits us to write certain fractions as simply one (whole) number.

Any whole number can be written as a fraction with a denominator of 1; so any fraction with a denominator of 1 can be simplified to the whole number in the numerator. Here are some examples:

$$7 = \frac{7}{1} \qquad\qquad 1 = \frac{1}{1} \qquad\qquad \frac{8}{1} = 8$$

From this it is clear that the reciprocal of a whole number is simply 1 over the number, e.g., the reciprocal of 10 is $\frac{1}{10}$.

---

[2] Unless the numerator is zero, because any fraction with a denominator of zero, such as $\frac{1}{0}$, is "not defined." Division by zero is "not defined" in mathematics.

The reciprocal of any whole number a is $\dfrac{1}{a}$[3].

3.1.5 Fractions with a value of 1

There is one more kind of simplifying, but it's really a special case of those fractions whose value is a whole number, such as we discussed above. We learned that any whole number can be written as a fraction with a denominator of 1.

If we do this with the number 1, we get

$$1 = \frac{1}{1}$$

Recalling from what we learned about equivalent fractions, the fraction $\frac{1}{1}$ could be re-written by multiplying the top and bottom by any number. But multiplying any number by 1 leaves the number unchanged, hence

$$\frac{1 \times 4}{1 \times 4} = \frac{4}{4} \qquad \frac{1 \times 3}{1 \times 3} = \frac{3}{3} \qquad \frac{1 \times 5}{1 \times 5} = \frac{5}{5} \qquad \frac{1 \times 231}{1 \times 231} = \frac{231}{231}$$

Since all these fractions began life as 1, they are all equal to 1. We see that any fraction in which the numerator is the same as the denominator must have the value 1.

Any fraction $\dfrac{a}{a}$, where the numerator is the same as the denominator, has the value 1.[4]

Looking closer at this fact reveals another aspect of fractions, one that is very significant for our understanding of them. Read on . . .

———————————

[3] Except zero. See above.

[4] $a \neq$ zero; division by zero is not defined. $\frac{0}{0} \neq 1$.

3.1.6 Fractions as division

We just learned that each of these fractions: $\frac{4}{4}$, $\frac{3}{3}$, $\frac{5}{5}$, $\frac{231}{231}$, has the same value, namely, 1.

In each of them, if we divide the numerator by the denominator we get 1, the simplified/reduced numerical value of the fraction.

Perhaps we should test this further, it may have some significance.

It's clearly true that if we divide the numerator of any whole-number valued fractions such as $\frac{7}{1}$, $\frac{3}{1}$, and $\frac{13}{1}$ by their denominators, 1, we get as a result the numerator, which is just the simplified whole-number value of the fraction. (Any number divided by 1 is unchanged.)

If we test our division hypothesis with any fraction in which the denominator is a factor of the numerator, we find that it holds up:

$$\frac{12}{3} \rightarrow 12 \div 3 = 4 \qquad \frac{35}{7} \rightarrow 35 \div 7 = 5 \qquad \frac{20}{4} \rightarrow 20 \div 4 = 5$$

So far, so good. It appears that by dividing the numerator of a fraction by its denominator, we arrive at the value of the fraction as a single number. *In fact, what this really means is that a fraction is simply another way of writing division!* The fraction bar merely replaces the division symbol, and we make sure to transfer the numbers around the " $\div$ " symbol left-to-right to top-to-bottom:

$$12 \div 3 \rightarrow \frac{12}{3} = 4 \qquad 27 \div 9 \rightarrow \frac{27}{9} = 3 \qquad 5 \div 8 \rightarrow \frac{5}{8}$$

Besides representing part of a whole, a fraction is used to convey division: the division of the numerator by the denominator.

Perhaps you noticed how in the last of the three examples above, we did not use an " $=$ " sign displaying the numerical value of the fraction $\frac{5}{8}$. That is because the fraction $\frac{5}{8}$ does not represent a whole number. It represents... *a fraction!*... a fractional part of a whole, such as 5 slices of pizza from a pie cut into eighths.

Just because 8 does not divide into 5 evenly does not mean that our hypothesis—that fractions mean division of numerator by denominator—was wrong. Hardly! You can divide 5 by 8. Do it on your calculator. The answer displayed is 0.625, a *decimal* number!

And indeed, this interpretation of fractions as division *could* lead us right into our next topic, *decimals*... except that it won't, because we still have a lot more to learn about fractions, namely, how to do arithmetic with them: adding, subtracting, multiplying and dividing.

But now that we do have a comprehensive understanding of fractions under our belts we are ready to learn how to use them with the basic arithmetic operations.

You-Try-It answers

1a. $\frac{6}{10}$; $\frac{9}{15}$; $\frac{12}{20}$ 1b. $\frac{4}{6}$; $\frac{6}{9}$; $\frac{8}{12}$ 2a. $\frac{17}{3}$ 2b. $\frac{80}{7}$

3a. $9\frac{1}{4}$ 3b. $16\frac{1}{2}$

Exercises 3.1

Find 3 equivalent fractions for each.

1. $\dfrac{3}{4}$

2. $\dfrac{2}{5}$

3. $\dfrac{3}{2}$

4. $\dfrac{7}{4}$

Reduce.

5. $\dfrac{12}{15}$

6. $\dfrac{5}{30}$

7. $\dfrac{48}{36}$

8. $\dfrac{27}{63}$

9. $\dfrac{420}{600}$

10. $\dfrac{2400}{45000}$

Simplify.

11. $\dfrac{8}{1}$

12. $\dfrac{35}{5}$

13. $\dfrac{0}{9}$

14. $\dfrac{253}{253}$

Convert to an improper fraction.

15. $2\dfrac{2}{3}$

16. $5\dfrac{3}{4}$

17. $8\dfrac{3}{5}$

18. $12\dfrac{5}{6}$

Convert to a mixed number.

19. $\dfrac{13}{3}$

20. $\dfrac{28}{9}$

21. $\dfrac{45}{7}$

22. $\dfrac{29}{13}$

Find the reciprocal.

23. $\dfrac{3}{4}$

24. $\dfrac{1}{9}$

25. 7

26. $\dfrac{0}{2}$

27. $\dfrac{14}{25}$

28. 1

3.2 Adding fractions

There are two types of addition of fractions, one is easy, the other hard.

The easy type is the addition of fractions having the same, or *like,* denominator.

3.2.1 Adding fractions with like denominators

Let's go back to our pizza pie cut into 8 slices. If you initially eat two slices, and then after others have eaten their full there is one slice left, being the good sport that you are, you finish off the last slice. So, you have eaten a total of three slices. The total amount of the pie you ate can be written mathematically like this:

$$\frac{2}{8} + \frac{1}{8} = \frac{3}{8}$$

Adding fractions with like denominators is indeed just this simple. You add up, or *count*, the numerators, while the denominator stays the same. Why should it change? What you are doing is counting how many slices of pizza you had. Each one is one-eighth, so if you eat three of them you've eaten three-eighths: $\frac{1}{8} + \frac{1}{8} + \frac{1}{8} = \frac{3}{8}$. Mathematically this is the same as writing

$$\frac{1 + 1 + 1}{8} = \frac{3}{8}$$

To add fractions with the same denominator, add the numerators and keep the denominator the same.

Now, after adding some fractions together it may turn out that the sum may be reducible. If it is you must reduce it. In general, when doing any operations with fractions we always reduce (or otherwise simplify) the final answer.

Consider these two examples of adding slices of pizza (eighths) that need to be reduced/simplified.

Example. If you at first eat three slices of pizza, and later eat one more, how much of the pie have you eaten?

$$\frac{3}{8} + \frac{1}{8} \;\longrightarrow\; \frac{3 + 1}{8} \;\longrightarrow\; \frac{4}{8}$$

But $\frac{4}{8}$ can be reduced. 4 is a factor of 8—and of course it is also a factor of itself, so to reduce we divide top and bottom by 4.

$$\frac{4 \div 4}{8 \div 4} = \frac{1}{2}$$

Because the numerator, 4, is a factor of the denominator, 8, a more direct way of doing the reducing in this case is to simply divide 4 into the denominator, 8, leaving 1 on top and 2 on the bottom:

$$\frac{4}{8} \longrightarrow \frac{\overset{1}{\cancel{4}}}{\underset{2}{\cancel{8}}} \longrightarrow \frac{1}{2}$$

You greedy pig! You ate half of the pie!

Example. You're like, *really* hungry for some pizza. Your buddies are over and you decide one pie is not going to be enough, so you order two pies. At the first sitting, each of you has 3 slices. A little later, you and the hungrier of your buddies each scarf down 2 more slices of the remaining pie. Your friends eventually take off leaving three slices in that second pie. Guess what? You eat them while watching a *Downton Abbey* marathon before going to bed.

How much pizza did you eat?

You first ate 3 slices, then 2, then 3, so:

$$\frac{3}{8} + \frac{2}{8} + \frac{3}{8} \longrightarrow \frac{3 + 2 + 3}{8} \longrightarrow \frac{8}{8}$$

But $\frac{8}{8}$ is a fraction with the same number on top and bottom, i.e., a fraction whose value is 1.

$$\frac{8}{8} = 1$$

Yes, I can't believe it either, but you ate a whole pie!

You-Try-It **1**

Add the fractions. Reduce or simplify the result if possible. If the result is an improper fraction, express it as a mixed number.

a) $\frac{1}{7} + \frac{3}{7} + \frac{2}{7}$
b) $\frac{3}{8} + \frac{1}{8} + \frac{2}{8}$
c) $\frac{1}{6} + \frac{5}{6} + \frac{3}{6}$

If you thought this was going too easy, you were right. We are now ready to learn how to add fractions with different denominators. This requires considerably more work.

3.2.2 Adding fractions with different denominators

Consider you bought two pizza pies but they were cut differently. One pie was cut into 8 slices, as before, but the other one was cut into 6 slices. Assume the pies are both the same size (otherwise we would be introducing an unnecessary complication).

If you ate 3 slices of the pie cut into eighths and one slice of the pie cut into sixths, how much pizza would you have eaten?

This is not simple to answer. It turns out we cannot add fractions unless they have the same denominator. Fractions with different denominators are called *unlike* fractions.

What this means is when we need to add fractions with different denominators, some or all of the fractions have to be changed to equivalent fractions so that *all the fractions have the same denominator*. We have to find a *common denominator* for the fractions before we can add them.

As a fraction addition problem, the above scenario would be written as $\frac{3}{8} + \frac{1}{6}$.

Adding fractions with different denominators is a 4-step process:

Step 1. Find the common denominator of the fractions.

Step 2. Change the fractions into equivalent fractions all having the common denominator as their denominator.

Step 3. Re-write the problem with the new fractions.

Step 4. Add the fractions and simplify/reduce the result.

Before we can attempt this we have to learn how to do *Step 1*, finding a common denominator.

3.2.2.1 LCDs and LCMs

We shall learn two different methods for finding common denominators.

The first method is usually sufficient for dealing with typical fractions, i.e., those we might come across in our day-day-day lives, with small denominators.

Using our pizza-slice example, we have two fractions, one with the denominator 6, the other 8.

We take successive multiples of the largest denominator and test it for divisibility by the other denominators.

So, first noting that 6 is not a factor of 8, we multiply 8 by 2, and see if that number is divisible by 6:

$$2 \times 8 = 16$$

Now we test if 6 divides evenly into 16. It doesn't. ($16 \div 6 = $ *Not a whole number*)

So now we go to the next multiple of 8: 3:

$$3 \times 8 = 24$$

Now we test if 6 divides evenly into 24. It does. ($24 \div 6 = 4$)

Therefore 24 is a *common denominator* for our *fractions* with denominators 6 and 8. Another way to say this is that **24 is a *common multiple* of the *numbers* 6 and 8**.

Notice, there is an even simpler way to find *a* common multiple of 6 and 8: simply multiply them together:

$$6 \times 8 = 48$$

We could use 48 for the denominator of the equivalent fractions we are going to make, but 24 is a smaller number, and so the equivalent fractions with *it* as a denominator will be smaller and easier to work with.

Twenty-four is the smallest, or *Least Common **Denominator*** (LCD) for our fractions $\frac{3}{8}$ and $\frac{1}{6}$; just as 24 is the *Least Common **Multiple*** (LCM) of 6 and 8.

When adding fractions with different denominators it is necessary to find *a* common denominator, but life will be a lot simpler if you find, and use, the ***Least*** Common Denominator, i.e., the *smallest* number that is divisible by the all denominators.

Definition. **Least Common Denominator** (LCD): Given two or more *fractions*, the smallest number that all the denominators will divide into evenly.

Definition. **Least Common Multiple** (LCM): Given two or more *numbers*, the smallest number that all the numbers will divide into evenly.

We see that these two mathematical concepts are really the same thing, it is only the context that differentiates them. When talking about LC***D***s, we're talking about fractions. When talking about LC***M***s, we're dealing with whole numbers (more often than not the denominators of fractions!).

For the purposes of our discussion here, I will refer to LCMs, until we get back to actually doing our fraction addition problem.

There are many methods for finding LCMs that do not rely on the trial-and-error nature of the *successive-multiples-of-the-largest-number* method. When trying to add three or more fractions with larger denominators, we soon realize its limitations. Let us look at one such method, with which we will find the same LCD for the fractions we are currently working with, $\frac{3}{8}$ and $\frac{1}{6}$ as above, i.e., another method to find the LCM of 6 and 8.

This method requires breaking down each number into its prime factorization. Prime numbers are numbers that have no factors (other than themselves and 1). The first ten prime numbers are: 1, 2, 3, 5, 7, 11, 13, 17, 19, 23.

We do this in a diagram which is sometimes called a tree-diagram, though it actually corresponds to an upside-down tree.

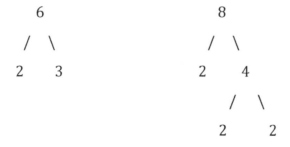

The last numbers on the *branches* are called the *leaves*. These numbers will be prime numbers after this branching process has been completed.

The first thing we do is note what numbers we have in our leaves. We see that 6 has two leaves: 2 and 3. We see that 8 has three leaves: 2, 2 and 2.

This tells us that the LCM of 6 and 8 will be some multiple of 2s and 3s. We just need to find out how many 2s and how many 3s.

First, the 2s. There is one 2 in 6. There are three 2s in 8. We take the greater of these numbers. So, our LCM will need *three* 2s.

Second, the 3s. Only 6 has a 3, and it has only one. Thus, our LCM needs *one* 3.

The LCM is then: $2 \times 2 \times 2 \times 3 = 24$.

The LCM of a group of numbers is formed from the product of their prime factors. For each prime factor appearing, we only take the ones from the number which has the greatest amount of them.

Example. Find the LCM of 8, 9, 12

If we used the *successive-multiples-of-the-largest-number* method, we would quickly grow frustrated:

$$2 \times 12 = 24 \qquad \text{Doesn't work for 9}$$
$$3 \times 12 = 36 \qquad \text{Doesn't work for 8}$$
$$4 \times 12 = 48 \qquad \text{Doesn't work for 9}$$
$$5 \times 12 = 60 \qquad \text{Doesn't work for 8 or 9}$$

We give up on this method and go to the second method:

```
     8                  9                 12
    / \                / \               / \
   2   4              3   3             3   4
      / \                               / \
     2   2                             2   2
```

Once again, the prime numbers that emerge in the leaves are 2 and 3. So once again our LCM will be some multiple of 2s and 3s. The question is: How many 2s and how many 3s?

First, the 2s:

> 8 has three 2s.
>
> 9 has no 2s.
>
> 12 has two 2s.

The greatest number of 2s is three, in 8. So, our LCM needs three 2s.

Second, the 3s:

> 8 has no 3s.
>
> 9 has two 3s.
>
> 12 has one 3.

The greatest number of 3s occurring in one of our numbers is two, within 9. So, our LCM needs two 3s.

$$LCM = 2 \times 2 \times 2 \times 3 \times 3 = 72$$

Well, looking back we see that if we would've continued with the *successive-multiples-of-the-largest-number* method we would've hit it with the very next multiple of twelve: six $(6 \times 12 = 72)$.

Oh well!

You-Try-It 2

Find the LCM of 9, 10 and 12.

We are now ready to add $\frac{3}{8} + \frac{1}{6}$.

Step 1.	Find the LCD.	The LCD is 24 (done above)
Step 2.	Re-write the fractions as equivalent fractions with the LCD as the denominator.	8 needs to be multiplied by 3 to become 24: $$\frac{3\times3}{8\times3} = \frac{9}{24}$$ 6 needs to be multiplied by 4 to become 24: $$\frac{1\times4}{6\times4} = \frac{4}{24}$$
Step 3.	Re-write the problem with the new fractions.	$$\frac{9}{24} + \frac{4}{24}$$
Step 4.	Add the fractions. No simplifying/reducing is possible.	$$\frac{9+4}{24}$$ $$\frac{13}{24}$$

Congratulations, you ate $\frac{13}{24}$ 'ths of a pizza!

You-Try-It 3

Add the fractions. Reduce/simplify if possible. If the result is an improper fraction, write it as a mixed number.

a) $\frac{1}{3} + \frac{8}{15}$

b) $\frac{3}{4} + \frac{1}{6} + \frac{2}{5}$

3.2.3 Adding mixed numbers

Now that we know how to add fractions, even with different denominators, we can add mixed numbers.

Recall that a mixed number such as $5\frac{2}{3}$ means 5 and $\frac{2}{3}$, or, 5 plus $\frac{2}{3}$. We use this knowledge to simplify adding mixed numbers.

Example. Add $12\frac{1}{3} + 5\frac{3}{4}$

Since $12\frac{1}{3}$ means $12 + \frac{1}{3}$, and $5\frac{3}{4}$ means $5 + \frac{3}{4}$, we can think of this addition of the two mixed numbers like this:

$$12 + \frac{1}{3} + 5 + \frac{3}{4}$$

Addition is *commutative*, which means that order doesn't matter; so we can write this sum as:

$$12 + 5 + \frac{1}{3} + \frac{3}{4}$$

The simplest way to do this is break it down into two separate sums; add the whole numbers together, then add the fractions. Then add the whole number to the fraction.

Step 1. $12 + 5 = 17$

Step 2. Add $\frac{1}{3} + \frac{3}{4}$

LCD is 12

$$\frac{1 \times 4}{3 \times 4} = \frac{4}{12}$$

$$\frac{3 \times 3}{4 \times 3} = \frac{9}{12}$$

$$\frac{4}{12} + \frac{9}{12} = \frac{13}{12}$$

The sum of the fractions is an improper fraction, i.e., greater than 1. Change it to a mixed number, then add it to 17, the sum of the whole numbers.

$$\frac{13}{12} = 1\frac{1}{12}$$

Step 3. Now do the final sum.

$$17 + 1\frac{1}{12} \;\rightarrow\; 17 + 1 + \frac{1}{12} \;\rightarrow\; 18\frac{1}{12}$$

You-Try-It 4

Add:

    a) $4\frac{3}{14} + 5\frac{1}{6}$ b) $9\frac{3}{5} + 8\frac{5}{6}$

You-Try-It answers

1a. $\frac{6}{7}$ 1b. $\frac{3}{4}$ 1c. $1\frac{1}{2}$ 2. 180

3a. $\frac{13}{15}$ 3b. $1\frac{19}{60}$ 4a. $9\frac{8}{21}$ 4b. $18\frac{13}{30}$

Exercises 3.2

Add. Reduce/simplify.

1. $\dfrac{4}{9} + \dfrac{1}{9} + \dfrac{2}{9}$

2. $\dfrac{3}{14} + \dfrac{5}{14} + \dfrac{2}{14}$

3. $\dfrac{2}{5} + \dfrac{4}{5} + \dfrac{9}{5}$

4. $\dfrac{3}{7} + \dfrac{5}{7} + \dfrac{2}{7}$

Find the LCM.

5. 6, 9

6. 10, 12

7. 3, 6, 10

8. 4, 8, 10

9. 5, 6, 15

10. 9, 12, 15

Find the LCD.

11. $\dfrac{2}{9}$, $\dfrac{5}{6}$

12. $\dfrac{9}{12}$, $\dfrac{5}{24}$

13. $\dfrac{3}{4}$, $\dfrac{1}{5}$, $\dfrac{2}{3}$

14. $\dfrac{1}{6}$, $\dfrac{1}{3}$, $\dfrac{4}{5}$

Add.

15. $\dfrac{1}{2} + \dfrac{3}{4}$

16. $\dfrac{2}{3} + \dfrac{5}{6}$

17. $\dfrac{1}{2} + \dfrac{1}{3}$

18. $\dfrac{3}{5} + \dfrac{1}{3}$

19. $\dfrac{1}{6} + \dfrac{1}{3} + \dfrac{4}{5}$

20. $\dfrac{2}{3} + \dfrac{3}{4} + \dfrac{1}{5}$

Add.

21. $3\frac{2}{7} + 5\frac{1}{7}$

22. $1\frac{2}{3} + 3\frac{1}{4}$

23. $4\frac{2}{5} + 6\frac{1}{3}$

24. $3\frac{3}{8} + 12\frac{1}{4}$

25. $7\frac{5}{6} + 2\frac{1}{3}$

26. $5\frac{5}{12} + 7\frac{8}{30}$

3.3 Subtracting Fractions

Subtracting fractions is done the same way we add them, by first ensuring they have the same denominator.

In this book we are going to avoid negative numbers, and so we ensure that any subtractions we do here will always be subtracting smaller quantities from larger ones.

Having said that, all the various complications we came across with addition, the need for common denominators, improper fractions and mixed numbers, etc., will also occur in subtraction. Also, in subtracting mixed numbers, we will again have to confront borrowing, to subtract a larger fraction from a smaller one.

We'll start off easy and work our way through the various complications.

Example. Subtract $\dfrac{11}{15} - \dfrac{8}{15}$

$$\frac{11-8}{15}$$

$$\frac{3}{15}$$

This reduces:

$$\frac{3 \div 3}{15 \div 3}$$

$$\frac{1}{5}$$

Example. Subtract $\dfrac{8}{9} - \dfrac{5}{6}$

Step 1.	Find the LCD.	9 x 2 = 18
		18 is divisible by 6
		The LCD is 18

Step 2.	Change the fractions to equivalent fractions with the denominator 18.	9 needs to be multiplied by 2 to become 18: $$\frac{8 \times 2}{9 \times 2} = \frac{16}{18}$$ 6 needs to be multiplied by 3: $$\frac{5 \times 3}{6 \times 3} = \frac{15}{18}$$
Step 3.	Re-write the problem with the new fractions.	$$\frac{16}{18} - \frac{15}{18}$$
Step 4.	Do the subtraction. No simplifying is possible.	$$\frac{16 - 15}{18}$$ $$\frac{1}{18}$$

Example. Subtract $16\frac{4}{9} - 12\frac{1}{2}$

If we attempted to do this problem in the same way we added mixed numbers we would run into difficulty because even though we are subtracting a smaller quantity from a larger, the fractional part that we are subtracting, $\frac{1}{2}$, is larger than $\frac{4}{9}$, the fractional part of the number we are subtracting from.

Step 1.	Find the LCD of the two fractions and change them to equivalent fractions having the LCD as the denominator	The LCD of $\frac{4}{9}$ and $\frac{1}{2}$ is 18 $$\frac{4 \times 2}{9 \times 2} = \frac{8}{18}$$ $$\frac{1 \times 9}{2 \times 9} = \frac{9}{18}$$
Step 2.	Write the subtraction in vertical format	$$16\frac{8}{18}$$ $$-12\frac{9}{18}$$

Step 3.	Borrow 1 from the 16 and write it as $\frac{18}{18}$. Add the two fractional parts together to form an improper fraction.	$16\frac{8}{18} \rightarrow 15\frac{8}{18} + \frac{18}{18} \rightarrow 15\frac{26}{18}$ $- 12\frac{9}{18}$
Step 4.	Subtract the whole numbers and subtract the fractions. Combine the resulting whole number and fraction into a mixed number. Simplify/reduce the fraction if necessary.	$15 - 12 = 3$ $\frac{26}{18} - \frac{9}{18} \rightarrow \frac{26-9}{18} \rightarrow \frac{17}{18}$ $3 + \frac{17}{18}$ $3\frac{17}{18}$

You-Try-It 1

Subtract:

a) $\frac{12}{15} - \frac{6}{15}$

b) $\frac{2}{3} - \frac{3}{10}$

c) $9\frac{1}{4} - 6\frac{3}{4}$

d) $18\frac{1}{3} - 11\frac{1}{2}$

You-Try-It answers

1a. $\frac{2}{5}$ 1b. $\frac{11}{30}$ 1c. $2\frac{1}{2}$ 1d. $6\frac{5}{6}$

Exercises 3.3

Subtract. Reduce/simplify.

1. $\dfrac{7}{9} - \dfrac{4}{9}$

2. $\dfrac{10}{11} - \dfrac{3}{11}$

3. $\dfrac{3}{4} - \dfrac{1}{2}$

4. $\dfrac{2}{3} - \dfrac{4}{9}$

5. $\dfrac{3}{5} - \dfrac{2}{10}$

6. $5\dfrac{5}{7} - 3\dfrac{1}{7}$

7. $8\dfrac{1}{4} - 6\dfrac{3}{4}$

8. $12\dfrac{2}{3} - 9\dfrac{1}{6}$

9. $5\frac{1}{4} - 3\frac{1}{2}$

10. $17\frac{10}{12} - 14\frac{2}{3}$

11. $9\frac{1}{5} - 8\frac{3}{10}$

12. $4\frac{5}{16} - 2\frac{2}{3}$

3.4 Multiplying fractions

The good news is that multiplying and dividing fractions is far simpler than adding and subtracting. Furthermore, once you learn how to multiply fractions, you are one trivial step away from division. These sections on multiplying and dividing will go a lot quicker than addition and subtraction.

First of all we should recall just what multiplication is: a compact way of doing repeated addition. For example, if we needed to add 5 to itself 7 times,

$$5 + 5 + 5 + 5 + 5 + 5 + 5$$

we use multiplication to represent this repeated sum as 5 times 7

$$5 \times 7$$

Multiplying fractions is no different. If we wanted to multiply $\frac{3}{8}$ by 5, what we mean is adding $\frac{3}{8}$ to itself 5 times. So,

$$\frac{3}{8} + \frac{3}{8} + \frac{3}{8} + \frac{3}{8} + \frac{3}{8}$$

is the same as

$$\frac{3}{8} \times 5$$

Let's first do the sum

$$\frac{3}{8} + \frac{3}{8} + \frac{3}{8} + \frac{3}{8} + \frac{3}{8}$$

$$\frac{3 + 3 + 3 + 3 + 3}{8}$$

$$\frac{15}{8}$$

So, this means that the product $\frac{3}{8} \times 5$ must, also equal $\frac{15}{8}$, that is

$$\frac{3}{8} \times 5 = \frac{15}{8}$$

If we write 5 as a fraction (numerator 5, denominator 1), this multiplication problem now looks like the product of 2 fractions

$$\frac{3}{8} \times \frac{5}{1} = \frac{15}{8}$$

Looking at this we would make an educated guess that the way to multiply fractions is to simply multiply the numerators and multiply the denominators like so,

$$\frac{3}{8} \times \frac{5}{1}$$

$$\frac{3 \times 5}{8 \times 1}$$

$$\frac{15}{8}$$

And guess what? Our educated guess would be right! To multiply together any number of fractions, we simply multiply the numerators across the top and then multiply the denominators across the bottom.

To multiply fractions, multiply the numerators and multiply the denominators.

$$\frac{a}{b} \times \frac{c}{d} \times \frac{e}{f} = \frac{a \times c \times e}{b \times d \times f}$$

Example. Multiply the fractions

a) $\frac{2}{3} \times \frac{4}{9}$

$$\frac{2 \times 4}{3 \times 9}$$

$$\frac{8}{27}$$

b) $\frac{3}{5} \times \frac{1}{6} \times \frac{10}{3}$

$$\frac{3 \times 1 \times 10}{5 \times 6 \times 3}$$

$$\frac{30}{90}$$

$\frac{30}{90}$ can be reduced:

$$\frac{1}{3}$$

Let's take a closer look at (b).

We see that the final answer reduced to $\frac{1}{3}$. We might wonder if there is shorter way to do this problem, one that eliminates the need to reduce at the end?

Indeed, there is—but before we demonstrate it, let's make clear that multiplying fractions as done in (b) above, i.e., just multiplying across the top and then multiplying across the bottom and **then** reducing, is a *perfectly fine way to multiply fractions.* Just make sure that you reduce to lowest terms after you've done the multiplying.

However, the preferred—and quite frankly *better*—way to do a problem like (b) above is to first simplify the fractions before doing the multiplications. This is best explained by a rather painful step-by-step breakdown of all the math going on "underneath," followed by the much shorter operational, step-by-step breakdown of how you actually do this.

Step 1.	$\dfrac{3}{5} \times \dfrac{1}{6} \times \dfrac{10}{3}$	Note that 3 is a factor of 6 and 5 is a factor of 10.
Step 2.	$\dfrac{3 \times 1 \times 10}{5 \times 6 \times 3}$	We can think of this product as one fraction.
Step 3.	$\dfrac{3 \times 1 \times 10}{3 \times 6 \times 5}$	We can reorder the numbers within the numerator and denominator because, like addition, multiplication is also *commutative*, meaning order doesn't matter (2×3 is the same as 3×2).
Step 4.	$\dfrac{3}{3} \times \dfrac{1}{6} \times \dfrac{10}{5}$	Now we can reverse what we did initially and consider this as the product of 3 separate fractions.
Step 5.	$1 \times \dfrac{1}{6} \times 2$	The fraction $\frac{3}{3}$ is just 1; and the fraction $\frac{10}{5}$ is just 2.
Step 6.	$1 \times 2 \times \dfrac{1}{6}$	Re-ordering.
Step 7.	$2 \times \dfrac{1}{6}$	Multiplying 1 by 2.

Step 8.	$\dfrac{2}{1} \times \dfrac{1}{6}$	Write 2 as a fraction.
Step 9.	$\dfrac{2 \times 1}{1 \times 6}$	Do the multiplying.
Step 10.	$\dfrac{2}{6}$	Reduce.
Step 11.	$\dfrac{1}{3}$	Final reduced answer.

Don't be scared by what looks like a complicated and lengthy set of steps. This is **not** how this is actually done. What is shown here is the "plumbing," if you will, underneath what is actually a *much shorter and simpler process*. Steps 2 through 5 are all done in one step, because you do not need to re-order to perform the multiplications. Likewise steps 6 through 10 are all done in one step.

The process will *actually* look like this:

Step 1.	$\dfrac{3}{5} \times \dfrac{1}{6} \times \dfrac{10}{3}$	Note 3 on the top and 3 on the bottom, and 5 is a factor of 10.
Step 2.	$\dfrac{\overset{1}{\cancel{3}}}{\underset{1}{\cancel{5}}} \times \dfrac{1}{6} \times \dfrac{\overset{2}{\cancel{10}}}{\underset{1}{\cancel{3}}}$	The $\frac{3}{3}$ is just 1, so cross out the 3s and replace them with 1s; the $\frac{10}{5}$ is just $\frac{2}{1}$, so cross out the 10 and 5 and replace them with 2 and 1 respectively.
Step 3.	$1 \times \dfrac{1}{\underset{3}{\cancel{6}}} \times \dfrac{\overset{1}{\cancel{2}}}{1}$	2 is a factor of 6; cross them out and replace them with 1 and 3 respectively; ($\frac{2}{6}$ reduces to $\frac{1}{3}$).
Step 4.	$1 \times \dfrac{1}{3} \times \dfrac{1}{1}$	This then is the actual product we have to find after doing all the reducing and simplifying. Once we do the multiplication now, it is guaranteed that the final answer will be in fully reduced form.

Step 5.	$\dfrac{1}{3}$	Final reduced answer.

Note a couple of things: Steps 2, 3 and 4 are all actually done in one place, with a lot of crossing out and writing of smaller numbers above and below the various numerators and denominators. Note also that an alternative exists where we could've first divided the numerator 3 in $\dfrac{3}{5}$ into the 6. If we had done this the remaining simplifying would've looked a little different, but the final result would be the same.

Here's how the problem would look if worked in one place.

$$\dfrac{\overset{1}{\cancel{3}}}{\underset{1}{\cancel{5}}} \times \dfrac{1}{\underset{3}{\cancel{6}}} \times \dfrac{\overset{\overset{1}{\cancel{2}}}{\cancel{10}}}{\underset{1}{\cancel{3}}}$$

Now that all the heavy lifting has been done, all we have to do is multiply $1 \times 1 \times 1$ across the top and $1 \times 3 \times 1$ across the bottom to arrive at the answer (in fully reduced form): $\dfrac{1}{3}$.

Whether or not you choose to simply/reduce *first* and then multiply, or multiply first and *then* reduce, is up to you. As long as you follow through correctly you will arrive at the same place: the correct answer, fully simplified.

You-Try-It 1

Multiply:

a) $\dfrac{4}{9} \times \dfrac{3}{5}$

b) $\dfrac{5}{2} \times \dfrac{3}{10} \times \dfrac{4}{9}$

3.4.1 Multiplying fractions, mixed numbers and whole numbers

Fractions tend to be rather selfish beasts. Seldom will they accommodate you if you are not one of them. If you want to interact with fractions, you have to enter *their* world and meet their demands.

If we need to multiply fractions with whole or mixed numbers, we first need to change them to improper fractions.

Example. Multiply

$$4 \times \frac{2}{3} \times 3\frac{1}{2}$$

In order to do the multiplication, we must change 4 and $3\frac{1}{2}$ to (improper) fractions.

$$4 = \frac{4}{1} \qquad\qquad\qquad 3\frac{1}{2} = \frac{7}{2}$$

Now we can do the multiplication.

$$\frac{4}{1} \times \frac{2}{3} \times \frac{7}{2}$$

I like to do my simplifying first:

$$\frac{4}{1} \times \frac{\overset{1}{\cancel{2}}}{3} \times \frac{7}{\underset{1}{\cancel{2}}}$$

$$\frac{4}{1} \times \frac{1}{3} \times \frac{7}{1}$$

Final result after multiplying across the top and bottom:

$$\frac{28}{3}$$

[You might optionally want to write this in its mixed number equivalent form: $9\frac{1}{3}$.]

You-Try-It 2

Multiply:

a) $2\frac{4}{5} \times \frac{3}{7} \times 10$

b) $\frac{2}{3} \times 3\frac{3}{4} \times 3$

You-Try-It answers

1a. $\frac{4}{15}$

1b. $\frac{1}{3}$

2a. 12

2b. $\frac{15}{2}$ *or* $7\frac{1}{2}$

Exercises 3.4

Multiply.

1. $\dfrac{1}{2} \times \dfrac{3}{5}$

2. $\dfrac{3}{4} \times \dfrac{3}{8}$

3. $\dfrac{3}{7} \times \dfrac{7}{3}$

4. $\dfrac{6}{5} \times \dfrac{7}{12}$

5. $\dfrac{3}{2} \times \dfrac{10}{9} \times \dfrac{4}{5}$

6. $\dfrac{3}{7} \times \dfrac{3}{5} \times \dfrac{7}{3}$

7. $2\dfrac{2}{3} \times 3\dfrac{3}{5}$

8. $1\dfrac{7}{11} \times 4\dfrac{1}{6}$

9. $1\dfrac{3}{4} \times 2 \times 2\dfrac{2}{3}$

10. $3\dfrac{5}{8} \times 2\dfrac{1}{3} \times 4\dfrac{4}{5}$

3.5 Dividing fractions

Here is some good news: once you know how to multiply fractions, you know 98% of what you need to know to divide them. Why not cut you a break and just give you the bottom line without all the razzmatazz for once? Here's the low-down:

To divide by a fraction, you multiply by its reciprocal.

Recall that you get the reciprocal of a fraction by flipping it. The reciprocal of $\frac{a}{b}$ is $\frac{b}{a}$.

So, to divide a number by $\frac{3}{4}$, for example, you would multiply it by $\frac{4}{3}$.

If the number you are dividing is a whole number or a mixed number, you know what you must do: convert it into a fraction. Then you form its reciprocal by flipping it. Then you do the multiplication.

Pretty straightforward, right? Let's try it.

Example. Divide

$$\frac{4}{5} \div \frac{7}{10}$$

First, form the reciprocal of $\frac{7}{10}$:

$$\frac{7}{10} \rightarrow \text{reciprocal} \rightarrow \frac{10}{7}$$

Now turn the division into a multiplication problem.

$$\frac{4}{5} \times \frac{10}{7}$$

Do the multiplication.

$$\frac{4}{\cancel{5}_1} \times \frac{\cancel{10}^2}{7}$$

$$\frac{4}{1} \times \frac{2}{7}$$

$$\frac{8}{7}$$

Example. Divide

$$\frac{5}{3} \div 2$$

The reciprocal of 2 is $\frac{1}{2}$.

$$\frac{5}{3} \times \frac{1}{2}$$

$$\frac{5}{6}$$

Example. After devouring most of two pizzas (each cut into 8 slices) with two buddies, there are five slices remaining. Suppose your two friends *insist* that the remaining slices be divided equally amongst the three of you.

(1) How many slices do each of you get? (You know that the answer is going to involve a fraction of a slice, since 5 is not evenly divisible by 3.)

(2) How much of a pie does this represent?

Answer:

(1) There are five slices to be divided three ways, so that is simply $5 \div 3$. Since this quotient is not a whole number, we write it as the fraction $\frac{5}{3}$ — after all, fractions mean division, right?

To answer the question in terms of slices, we need this improper fraction as a mixed number.

$$\frac{5}{3} \rightarrow 1\frac{2}{3}$$

Each of you gets $1\frac{2}{3}$ slices. Have fun fighting over who has to get the $\frac{2}{3}$ in two separate one-third pieces.

(2) We want to know what part of one whole pie corresponds to one-third of 5 slices. This is a fraction division problem. We need to divide the $\frac{5}{8}$ of the pie by 3.

$$\frac{5}{8} \div 3$$

The reciprocal of 3 is $\frac{1}{3}$, so the problem becomes

$$\frac{5}{8} \times \frac{1}{3}$$

$$\frac{5}{24}$$

Now, admittedly, I have no idea why anyone would want to know something like this, but it gave us a chance to do some fraction division with a whole number. Cheers!

You-Try-It 1

Divide:

a) $\frac{8}{25} \div \frac{2}{5}$

b) $12 \div \frac{2}{3}$

c) $5\frac{2}{3} \div 2$

d) $\frac{3}{4} \div 3\frac{1}{2}$

You-Try-It answers

1a. $\frac{4}{5}$

1b. 18

1c. $\frac{17}{6}$ *or* $2\frac{5}{6}$

1d. $\frac{3}{14}$

Exercises 3.5

Divide.

1. $\dfrac{2}{3} \div \dfrac{1}{6}$

2. $\dfrac{3}{4} \div 2$

3. $\dfrac{5}{3} \div 3$

4. $\dfrac{5}{3} \div \dfrac{1}{3}$

5. $\dfrac{3}{4} \div \dfrac{1}{2}$

6. $\dfrac{4}{5} \div \dfrac{3}{2}$

7. $\dfrac{5}{4} \div \dfrac{5}{2}$

8. $3\dfrac{1}{2} \div \dfrac{2}{3}$

9. $\dfrac{4}{9} \div 3\dfrac{1}{3}$

10. $5\dfrac{3}{4} \div 2\dfrac{1}{2}$

Chapter 4 Decimals

4.1 Basic concepts

In order to explain decimals we first remind you of the place-value system from Chapter 2, and then show how it is extended to place values corresponding to quantities less than 1.

Our number system is a decimal, or base 10, system. This means that there must be ten different numerals (or *digits*): 0,1,2,3,4,5,6,7,8,9. All numbers can be made up of the numerals in various combinations.

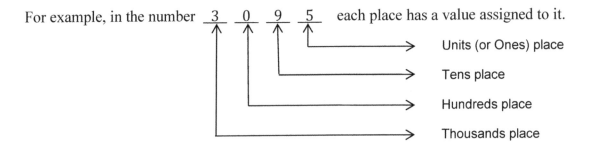 Each position, called a *place value*, can include any of the 10 numerals.

For example, in the number 3 0 9 5 each place has a value assigned to it.

Units (or Ones) place

Tens place

Hundreds place

Thousands place

Thus, the number 3095 means

	5×1	=	5
+	9×10	=	90
+	0×100	=	0
+	3×1000	=	3000
	Total		3095

For numbers smaller than 1, we can either use fractions or extend the base-10 decimal system beyond (to the right of) the smallest place-value, the ones (or units) place. For example: 3095.3

The decimal point separates the whole-number part from the *fractional decimal part*. The value of the fractional decimal part is less than 1, regardless of how many digits it contains.

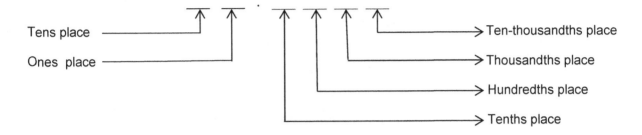

There is no limit to the place-values both to the left and to the right; however, our naming conventions are limited, generally speaking, to the trillions place on the left side, and, though we might say "trillionths" for the 13th place to the right of the decimal point, it is common practice to stop using names for the fractional place-values beyond *thousandths*, and identify them by number, as in "thirteenth decimal place."

(Once we start dealing with very tiny numbers, as well as very large numbers like in astronomy, we prefer to use something called *Scientific Notation*.)

Let's disassemble a fractional decimal number just as we did for the number 3095.

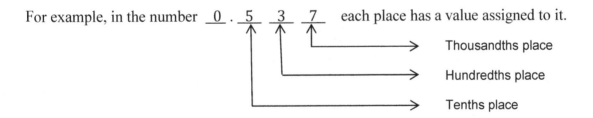

Note it is customary to write a leading zero (0.×××) with a fractional decimal number, i.e., when there is no whole-number part.

Thus, the number 0.537 means

$$
\begin{array}{rcl}
0 \times 1 & = & 0 \\[2mm]
+ \quad 5 \times \dfrac{1}{10} & = & \dfrac{5}{10} \\[4mm]
+ \quad 3 \times \dfrac{1}{100} & = & \dfrac{3}{100} \\[4mm]
+ \quad 7 \times \dfrac{1}{1000} & = & \dfrac{7}{1000} \\[3mm]
\hline
\text{Total} & = & \dfrac{5}{10} + \dfrac{3}{100} + \dfrac{7}{1000}
\end{array}
$$

But, once again the demanding nature of fractions reveals itself, and in order to see what this decimal number represents as a fraction we have to perform fraction addition with unlike denominators.

The LCD is 1000, so $\frac{5}{10}$ and $\frac{3}{100}$ must both be changed into thousandths.

$$
\frac{5}{10} \times \frac{100}{100} = \frac{500}{1000} \qquad\qquad \frac{3}{100} \times \frac{10}{10} = \frac{30}{1000}
$$

$$
\frac{500}{1000} + \frac{30}{1000} + \frac{7}{1000}
$$

$$
\frac{500 + 30 + 7}{1000}
$$

$$
\frac{537}{1000}
$$

But this number was arrived at by disassembling the decimal number 0.537!

We see that the fraction $\frac{537}{1000}$ is exactly the same as the decimal number 0.537.

This should not be a surprise—in fact it would be a great surprise indeed if it turned out to be any other number, because 0.537 means *537 thousandths*. The third place to the right of the decimal point is *thousandths*, and we have 537 of them!

Similarly

$$0.35 \text{ is } \frac{35}{100}; \text{ or } 35 \text{ hundredths}$$

$$0.102 \text{ is } \frac{102}{1000}; \text{ or } 102 \text{ thousandths}$$

$$0.5604 \text{ is } \frac{5604}{10000}; \text{ or } 5{,}604 \text{ ten-thousandths}$$

4.1.1 Rounding

If we are dealing with a number like 2.54706, we may not need all the precision of the number. After all, six hundred-thousandths ($\frac{6}{100000}$) is really a very tiny number. If we need to do some calculation with this number, we don't need to know the exact amount, we just need to find a "ballpark" figure.

It is very common to *round* decimal numbers. When a number is *rounded* it is simplified to some less precise number. If it were only necessary to know 2.54706 to the tenths place position, it would be rounded to 2.5, because it is closer to 2.5 than it is to 2.6.

These are the steps in rounding:

Step 1. Identify the digit in the place value you are rounding to.

Step 2. Identify the digit to its immediate right. This is called the *critical digit.*

Step 3. If the critical digit is 5 or greater, you round *up*. If the critical digit is less than 5, you round *down*.

Step 4. If you are rounding up, you add one to the digit in the place you are rounding to. If the digit there is a 9, you change it to 0 and bump the digit to its left by 1. This 'bumping' process can occur more than once.

If you are rounding down, you leave the digit in the place you are rounding to unchanged.

Step 5. Drop all digits to the right of the place you have rounded to.

A rounding position may be identified either by its place-value name (eg. "hundredths") or by its numerical order, as in "3 decimal places," which refers to the third position after the decimal point, which would be the thousandths place.

Let's do some examples to show exactly how this works.

Example. Round 2.54706 to the nearest tenth

Step 1.	Underline the digit in the tenths place.	2.5̲4706
Step 2.	Identify the critical digit, the digit to its immediate right: 4.	2.5̲4706 *Critical digit* = 4
Step 3.	The critical digit is less than 5, so we round down.	2.5̲4706 Round down
Step 4.	Leave the 5 unchanged; drop the rest of the digits.	2.5

Example. Round 2.54706 to the nearest hundredth

Step 1.	Underline the digit in the hundredths place.	2.54̲706
Step 2.	Identify the critical digit, the digit to its immediate right: 7.	2.54̲706 *Critical digit* = 7
Step 3.	The critical digit is greater than 5, so we round up.	2.54̲706 Round up
Step 4.	Add 1 to the 4, bumping it to 5; drop the rest of the digits.	2.55

Example. Round 2.09952 to 2 decimal places

Step 1.	Underline the digit in the 2nd decimal (hundredths) place.	2.09̲952

Step 2.	Identify the critical digit, the digit to its immediate right: 9.	2.0<u>9</u>952 *Critical digit* = 9
Step 3.	The critical digit is greater than 5, so we round up.	2.0<u>9</u>952 Round up
Step 4.	Adding 1 to the 9 gives 10. Write 0 in the hundredths place and bump the digit in the tenths place: 0; 0→1. Do not drop the zero in the hundredths place! You are rounding to that place and so you must have a digit there—even if it's zero! Drop the remaining digits.	2.10

Example. Round 2.09952 to the nearest thousandth

Step 1.	Underline the digit in the thousandths place.	2.09<u>9</u>52
Step 2.	Identify the critical digit, the digit to its immediate right: 5.	2.09<u>9</u>52 *Critical digit* = 5
Step 3.	The critical digit is 5, so we round up.	2.09<u>9</u>52 Round up

Step 4.	Adding 1 to the 9 gives 10. Write 0 in the thousandths place and bump the digit in the hundredths place: add 1 to the 9.	2.100
	This gives us 10 in the hundredths place. Write 0 in the hundredths place and bump the digit in the tenths place: $0 \rightarrow 1$.	
	Write 1 in the tenths place.	
	Do not drop the zero in the hundredths place and do not drop the zero in the thousandths place! You are rounding to that place and so you must have a digit there— even if it's zero!	
	Drop the remaining digits.	

You-Try-It 1

Round 1.093599 to:

a) the nearest tenth

b) 4 decimal places

c) the nearest thousandth

d) the nearest hundredth

e) 5 decimal places

f) the nearest one

4.1.2   Changing decimals to fractions

Here is a very easy way to write any decimal number as a fraction.

Example.   Write 0.432117 as a fraction

Step 1.	Count the number of place-values to the right of the decimal point.	Six
Step 2.	Write a fraction with a '1' followed by this number of zeros in the denominator.	$\dfrac{}{1000000}$
Step 3.	Write the numerals in the decimal in the numerator.	$\dfrac{432117}{1000000}$

Note, there is no guarantee that your fraction will be in lowest terms. If it can be reduced, as always, reduce it.

If you need to write a decimal number that also has a whole-number part as a fraction, this method will still work.

Example.   Write 23.054 as a fraction

Step 1.	Count the number of place-values to the right of the decimal point.	Three
Step 2.	Write a fraction with a '1' followed by this number of zeros in the denominator.	$\dfrac{}{1000}$
Step 3.	Write all the numerals in the decimal number in the numerator, ignoring the decimal point.	$\dfrac{23054}{1000}$
Step 4.	Reduce.	$\dfrac{23054 \div 2}{1000 \div 2} = \dfrac{11527}{500}$

If you prefer to see 23.054 as a mixed number:

Step 1.	Separate the whole-number part from the decimal part.	$23.054 = 23 + .054$
Step 2.	Count the number of place-values to the right of the decimal point.	Three
Step 3.	Write a fraction with a '1' followed by this number of zeros in the denominator.	$\dfrac{}{1000}$
Step 4.	Write all the numerals in the decimal part in the numerator. No need to write the leading zero.	$\dfrac{54}{1000}$
Step 5.	Reduce.	$\dfrac{54 \div 2}{1000 \div 2} = \dfrac{27}{500}$
Step 6.	Reunite the fraction while its whole-number part, 23.	$23 + \dfrac{27}{500} = 23\dfrac{27}{500}$

You-Try-It 2

Write as a fraction, proper or improper. Reduce if possible.

a) 0.6397   b) 12.605

Write as a mixed number. Reduce if possible.

c) 3.018   d) 43.56

You-Try-It answers

1a. 1.1   1b. 1.0936   1c. 1.094   1d. 1.09

1e. 1.09360   1f. 1

2a. $\frac{6397}{10000}$ 2b. $\frac{2521}{200}$ 2c. $3\frac{9}{500}$ 2d. $43\frac{14}{25}$

Exercises 4.1

Round 5.10963 to the indicated place-value.

1. the nearest tenth

2. two decimal places

3. four decimal places

4. the nearest thousandth

5. the nearest one

6. the nearest ten

(Always reduce/simplify fractions)

Write as a fraction.

7. 0.12

8. 0.375

9. 0.014

10. 0.09

Write as an improper fraction.

11. 1.25   12. 1.05

13. 4.08   14. 10.1

Write as a mixed number.

15. 1.4   16. 5.05

17. 3.125   18. 47.065

4.2 Adding and subtracting decimal numbers

4.2.1 Addition

The only consideration forced on us by the decimal point is that you respect his position. You shall not ignore him. You shall not misplace him.

To add decimal numbers you write them vertically aligned *by their decimal points*. Refer to the Standard Addition algorithm in 2.2.1 if necessary.

Example. Add 13.8702 to 7.518

$$
\begin{array}{r}
1\ \ \ 1\ \ \ \ \ \ \ \ \ \ \ \ \\
1\ 3\ .\ 8\ 7\ 0\ 2\\
+\ \ \ \ \ 7\ .\ 5\ 1\ 8\ \ \\
\hline
2\ 1\ .\ 3\ 8\ 8\ 2
\end{array}
$$

You-Try-It 1

Add:

a) $605.834 + 92.513$

b) $12.007632 + 0.0376$

4.2.2 Subtraction

We subtract decimal numbers similar to how we subtract whole numbers, with the proviso that we respect the position of the decimal point. We vertically align the numbers by their decimal points. If you need a refresher on subtracting, refer to the Standard Subtraction algorithm in 2.3.1.

Example. Subtract 87.043 from 231.67

We will need to write a zero in the thousandths place of 231.67 since the number we are subtracting from it extends to the thousandths position.

$$
\begin{array}{r}
\ \ 1\ \ ^12\ \ 1\ \ \ \ \ \ \ \ 6\ \ _1\\
\ \ \cancel{2}\ \ \cancel{3}\ \ 1\ .\ 6\ \cancel{7}\ 0\\
+\ \ \ \ \ 8\ \ 7\ .\ 0\ 4\ 3\\
\hline
1\ \ 4\ \ 4\ .\ 6\ 2\ 7
\end{array}
$$

You-Try-It 2

Subtract:

    a) 836.408 − 68.535   b) 78.08239 − 7.072967

You-Try-It answers

1a. 698.347   1b. 12.045232   2a. 767.873   2b. 71.009423

Exercises 4.2

Add.

1. $15.6 + 8.1$

2. $9.18 + 0.6$

3. $12.07 + 8.52$

4. $8.502 + 11.78$

5. $0.105 + 0.998 + 0.64$

6. $15.0843 + 23.157 + 0.82$

Subtract.

7. $15.84 - 11.63$

8. $12.94 - 10.57$

9. $6.1 - 4.39$

10. $3 - 0.054$

11. $4.02 - 1.367$

12. $12.0524 - 8.43$

4.3 Multiplying decimal numbers

Unlike addition and subtraction, we format decimal numbers for multiplication just as we do whole numbers: right-aligned. We ignore the decimal points while doing the multiplication operations, and only take account of their positions in the answer at the end.

Example. Multiply 15.4×3.7

Step 1.	Write in vertical format with the number with the least number of digits on the bottom. Keep the decimal points in place.	$$\begin{array}{r} 1\ \ 5\ .4 \\ \times \qquad 3\ .7 \\ \hline \end{array}$$
Step 2.	Do the multiplication; *ignore the decimal points*.	$$\begin{array}{r} 1\ \ 5\ .4 \\ \times \qquad 3\ .7 \\ \hline 1\ \ 0\ \ 7\ \ 8 \\ 4\ \ 6\ \ 2 \quad \\ \hline 5\ \ 6\ \ 9\ \ 8 \end{array}$$
Step 3.	Count the number of places to the right of the decimal point in both of the numbers being multiplied.	There are 2 decimal places total in 15.4 *and* 3.7.
Step 4.	Count that number of places right-to-left in the answer and write in the decimal point.	5698 $\leftarrow$ (2 places) 56.98

Example. Multiply 57.042×2.43

Step 1.	Write in vertical format with the number with the least number of digits on the bottom. Keep the decimal points in place.	$$\begin{array}{r} 5\ \ 7\ .0\ \ 4\ \ 2 \\ \times \qquad 2\ .4\ \ 3 \\ \hline \end{array}$$

Step 2.	Do the multiplication; *ignore the decimal points*.	

$$
\begin{array}{r}
5\ 7\ .0\ 4\ 2 \\
\times\ \ \ \ \ \ 2\ .4\ 3 \\
\hline
1\ 7\ 1\ 1\ 2\ 6 \\
2\ 2\ 8\ 1\ 6\ 8\ \ \ \\
1\ 1\ 4\ 0\ 8\ 4\ \ \ \ \ \\
\hline
1\ 3\ 8\ 6\ 1\ 2\ 0\ 6
\end{array}
$$

Step 3.	Count the number of places to the right of the decimal point in both of the numbers being multiplied.	There are 5 decimal places total in 57.042 *and* 2.43.
Step 4.	Count that number of places right-to-left in the answer and write in the decimal point.	13861206 ← (5 places) 138.61206

You-Try-It 1

Multiply:

    a) 4.6 × 85.2   b) 3.62 × 42.015

4.3.1 Multiplying decimal numbers by powers of 10

Powers of 10 are what you get when 10 is multiplied by itself repeatedly.

Ten multiplied by 10 is 100. This is the 2nd power of 10. Note there are 2 zeros in 100. Ten multiplied by itself 3 times, $10 \times 10 \times 10 = 1000$. 1,000 is the 3rd power of 10. There are 3 zeros in 1,000.

If a whole number is multiplied by a power of 10, you just add zeros corresponding to the number of zeros in the power of 10.

$$13 \times 100 = 1300$$

$$412 \times 10,000 = 4,120,000$$

What about multiplying decimal numbers by 10, or 100, or any power of 10, i.e., 1000, 10000, etc.?

If we multiply 0.6 by 10, we get 6. To multiply 0.6 by 10, you multiply 6 by 10, which is 60, then note there is only 1 digit to the right of a decimal point (the 6 in 0.6), so you count 1 place to the left from 60 and drop in the decimal point: 6.0, which is simply 6.

If we wanted to multiply 0.6 by 100, the multiplication would yield 600, and the 1 digit to the right of the decimal in 0.6 means that we drop a decimal point one place back from the right-most digit in 600: 60.0 . Which is just 60. So 0.6 times 100 is 60.

To multiply 0.6 by 10 we just move the decimal point 1 places to the right. Done.

To multiply 0.6 by 100 we just move the decimal point 2 places to the right, adding the necessary additional zero. Done.

To multiply a decimal number by a power of 10 you move the decimal point to the right the number of places corresponding to the number of zeros in the power of 10, adding zeros if necessary.

You-Try-It 2

Multiply:

    a) 4.635×100 b) 3.62×10

    c) 0.015×100 d) 14.001×1000

You-Try-It answers

1a. 391.92   1b. 152.0943

2a. 463.5   2b. 36.2   2c. 1.5   2d. 14,001

Exercises 4.3

Multiply.

1. 4.5 × 8.2

2. 13.4 × 8.6

3. 2.45 × 0.18

4. 3.14 × 2.16

5. 25.3 × 5.062

6. 24.513 × 3.43

Multiply mentally by moving the decimal point.

7. 3.52 × 10

8. 14.065 × 1,000

9. 8.54 × 100

10. 0.006809 × 10,000

4.4 Dividing decimal numbers

Today, few people do division by hand. The calculator is just too convenient. Why bother going through the laborious process of doing long division when technology provides us with the means to do it with just a few mouse-clicks or button-presses?

We're not going to enter that philosophical debate here. We'll just show you how it's done and then leave it to you to decide whether or not you want to perform division by hand or take the coward's way out.

4.4.1 Decimal long division

Recall that the numbers in a division problem have names. If we divide 12 by 3 — i.e., $12 \div 3$ —12 is called the *dividend* and 3 is the *divisor*. The answer to a division problem is called the *quotient*, which would be 4 in this case.

Example. Divide 1381.941 by 24.3

Step 1.	Write in standard long-division format. Keep the decimal points in place, initially.	$2\ 4\ .3\ \overline{\rvert 1\ \ 3\ \ 8\ \ 1\ \ .9\ \ 4\ \ 1}$
Step 2.	Move the decimal point in the divisor to make it a whole number, noting the number of places the decimal point moves.	$24.3 \rightarrow 243$ The decimal point is moved one place.
Step 3.	Move the decimal point in the dividend (to the right) the same number of places.	We move the decimal point one place in the dividend: $1381.941 \rightarrow 13819.41$
Step 4.	Rewrite the division problem. Copy the decimal point directly above into the quotient area on top.	$2\ 4\ 3\ \overline{\rvert 1\ \ 3\ \ 8\ \ 1\ \ 9\ \ .4\ \ 1}$

Step 5.	Do the division. The decimal point in the quotient area is in the right place—just pretend it isn't there while doing the long division steps.	

```
                            5  6 .8  7
         2  4  3 │1  3  8  1  9 .4  1
                   1  2  1  5
                   ─────────
                      1  6  6  9
                      1  4  5  8
                      ─────────
                         2  1  1  4
                         1  9  4  4
                         ─────────
                            1  7  0  1
                            1  7  0  1
                            ─────────
                                     0
```

You-Try-It 1

Divide:

    a) $18.011 \div 8.3$ b) $50.142 \div 2.74$

4.4.2 Dividing decimal numbers by powers of 10

As we learned in 4.3.1, to multiply a decimal number by a power of 10 (10, 100, 1000, etc.) we just move the decimal point to the right (the direction that makes the number bigger) the corresponding number of places—one place for each zero in the power of 10.

To divide a number by a power of 10 we do just the opposite: move the decimal point to the *left* (the direction that makes the number smaller) the number of places corresponding to the number of zeros in the power of 10.

Example. Divide. $57.042 \div 1000$

Step 1.	There are three zeros in 1,000 so we move the decimal point three places to the left, inserting the necessary zero in the tenths place.	.057042
Step 2.	Write a leading zero in the ones place.	0.057042

You-Try-It 2

Divide:

 a) $463.5 \div 100$

 b) $9.87 \div 10$

 c) $0.043 \div 1000$

 d) $64.005 \div 1000$

4.4.3 Changing fractions to decimals

There is one simple rule for changing a fraction into its decimal form:

Do the division.

Example. Write $\frac{3}{4}$ as a decimal

Step 1.	Divide 3 by 4	
		$\begin{array}{r} .7\ 5 \\ 4\ \overline{\smash{\big)}\ 3\ .0\ 0} \\ \underline{2\ 8} \\ 2\ 0 \\ \underline{2\ 0} \\ 0 \end{array}$
Step 2.	Done. The result is 0.75	$\frac{3}{4} = 0.75$

Example. Write $\frac{2}{3}$ as a decimal

Step 1.	Divide 2 by 3	
		$\begin{array}{r} .6\ 6\ 6 \\ 3\ \overline{\smash{\big)}\ 2\ .0\ 0} \\ \underline{1\ 8} \\ 2\ 0 \\ \underline{1\ 8} \\ 2\ 0 \end{array}$
Step 2.	We see this is a repeating decimal. The 6s go on forever. A repeating pattern in a decimal is indicated by placing a bar over the digit (or digits) that repeat.	$\frac{2}{3} = 0.\overline{6}$
Step 3.	Sometimes we are asked to round a repeating decimal.	$\frac{2}{3} \approx 0.7$

Note, a repeating decimal can have any number of digits in its pattern. The decimal form of the fraction $\frac{1}{7}$ has a repeating pattern six digits long:

$$0.\overline{1\ 4\ 2\ 8\ 5\ 7}$$

You-Try-It **3**

Change to a decimal:

a) $\frac{4}{5}$ b) $\frac{3}{8}$

c) $\frac{5}{9}$ d) $\frac{3}{7}$

You-Try-It answers

1a. 2.17   1b. 18.3

2a. 4.635   2b. 0.987   2c. 0.000043   2d. 0.064005

3a. 0.8   3b. 0.375   3c. $0.\overline{5}$ 3d. $0.\overline{428571}$

Exercises 4.4

Divide.

1. 28.9 ÷ 3.4

2. 122.04 ÷ 5.4

3. 12.994 ÷ 0.73

4. 152.75 ÷ 6.5

Divide mentally by moving the decimal point.

5. 31.6 ÷ 100

6. 82.04 ÷ 10

7. 5073.8 ÷ 1000

8. 14.903 ÷ 1000

Change to a decimal number.

9. $\dfrac{1}{4}$

10. $\dfrac{11}{10}$

11. $\dfrac{1}{8}$

12. $\dfrac{3}{5}$

13. $\dfrac{5}{8}$

14. $\dfrac{5}{2}$

15. $\dfrac{1}{3}$

16. $\dfrac{5}{6}$

17. $\dfrac{4}{9}$

18. $\dfrac{1}{7}$

Chapter 5 Percent

Percents are yet another way to represent numbers. You might be wondering at this point: there's whole numbers, fractions, decimals—why do we need another form for writing a number?

Often we need to compare two or more numbers. We might want to compare the poll numbers of several candidates for political office. Or we may be preparing to buy a car and want to figure out how much more it will cost us to drive the car with lower gas mileage.

When we compare two numbers for reasons like this we usually use a *ratio*. A ratio is a comparison of two numbers written as a fraction (more formally speaking: as a quotient).

Let's say for some reason we needed to compare two oddball numbers such as 337 and 1679.

If we used a ratio to compare these, the fraction would be $\frac{337}{1679}$. Most people would not be able to get a good mental picture of what this quantity represents, what the actual *size* of this number is.

This is where percents come in handy. If this fraction is written as a percent, it is very close to 20%.

Twenty is a number that most people can wrap their heads around. So, once we explain what the "%" symbol means, the value of using percents will be apparent.

5.1 Percent definition

The word "percent" is of Latin origin, coming from ancient Rome. Literally, "per centum" means "by hundred." So, percent clearly has something to do with the number 100.

The concept is actually quite simple. To enter the world of percent we just imagine that the number 1—that is, one whole thing of any sort—becomes magnified to100.

$$1 \rightarrow 100\%$$

... and we attach the "%" to the number now in percent form. (Do NOT forget to attach the % symbol!)

A percent is a measure of a number considered as part of 100. For example 3% means the quantity represented by 3 parts out of 100, which is the same as the fraction $\frac{3}{100}$. Here is a visual representation of 3%:

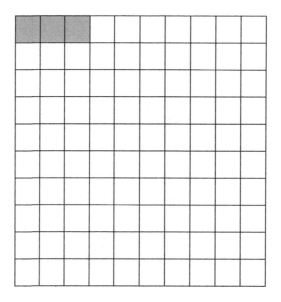

Likewise, 20% means 20 parts out of 100, which as a fraction is $\frac{20}{100}$ (which reduces to $\frac{1}{5}$).

Essentially, to change a number to its percent form, we just multiply it by 100.

Often when percents are taught a variety of rules are given to show how to change numbers in their various forms—whole numbers, fractions (including improper fractions and mixed numbers), and decimal numbers—to percents; and then more rules to reverse the process (i.e., to change numbers in percent form back into whole numbers, fractions or decimals).

Actually, to do these conversions **there are really only two rules**: a rule to change a number (in whatever form) to a percent, and a rule to change a percent into a number (in whatever form).

Here they are:

Rule 1: To change a number into a percent:

Multiply the number by 100

Rule 2: To change a percent to a number:

Divide the percent by 100

If you ever get bogged down in a percent problem, go back to these two rules. All percent problems emanate from one of these two rules. Get your bearings. Decide what it is you are trying to do—trying to find a number, or trying to find a percent? —and then start with the appropriate rule. At some point you will either have to do multiplication by 100 or division by 100...

Now, naturally, the devil is in the detail. The calculations will be different depending on whether you are trying to change $\frac{337}{1679}$, or 0.54, or $2\frac{3}{5}$ to a percent, but the really important thing is what Rule 1 tells you: You have to multiply it by 100!

How you do the multiplication depends, of course, on the form of the number. If you are changing $\frac{337}{1679}$ to a percent, you are dealing with fractions. What did we say about fractions? They are stubborn beasts. It's my way or the highway. To work with fractions, you must enter *their* world. They will not accommodate you until you are willing to play by their rules.

So, if you needed to change $\frac{337}{1679}$, you poor thing you, you will have to do fraction multiplication with this ugly fraction.

Okay, enough preamble, we're now ready to learn how to implement Rule 1, that is, how to change a number in any form into a percent.

5.2 How to change a number to a percent

We'll break this down by the three types of numbers we are familiar with: whole numbers, fractions (and their variants), and decimals.

5.2.1 Changing a whole number to a percent

Recall that we *define* the concept of percent by establishing that 1 is 100%. We can think of that as multiplying 1 by 100 (following Rule 1), but there is actually a much simpler way. The number 100 is just 1 with two 0's added. This will hold true for any whole number, not just 1.

To multiply a number by 100 you can just add two zeroes. So,

$$2 = 200\%$$

$$3 = 300\%$$

$$. . .$$

$$12 = 1200\%$$

Yes, it's just that simple.

You-Try-It 1

Change to a percent:

a) 5   b) 30

5.2.2 Changing a fraction to a percent

Recall fractions have three sub-classifications:

Proper fractions: Fractions whose value is less than 1, whose numerator is less than the denominator, like or $\frac{337}{1679}$, or $\frac{1}{2}$.

Improper fractions: Fractions whose value is greater than 1, whose numerator is greater than the denominator, such as $\frac{7}{3}$.

Mixed numbers: Numbers that are composed of a whole number part and fraction part, like $2\frac{3}{5}$.

A mixed number can always be represented as an improper fraction, as its value must be greater than 1 (it has a whole-number part).

5.2.2.1 Changing a proper fraction to a percent

There is nothing beyond Rule 1: Multiply the fraction by 100. The only thing required is that you know how to multiply fractions. That is not grounds for any new rule. Multiplying fractions is one of the skills you must know to do arithmetic with fractions. We learned how to do it in 3.4.

Let's do a couple:

Example. Change $\frac{3}{4}$ to a percent

Step 1.	Multiply the fraction by 100	$\frac{3}{4} \times 100$
Step 2.	Write 100 as a fraction	$\frac{3}{4} \times \frac{100}{1}$
Step 3.	Simplify before doing the multiplication. Four goes into 100 twenty-five times.	$\frac{3}{\overset{}{\underset{1}{4}}} \times \frac{\overset{25}{\cancel{100}}}{1}$
Step 4.	Multiply across top and bottom	$\frac{3 \times 25}{1 \times 1} = \frac{75}{1}$
Step 5.	Bottom is 1, so the result is just the whole number 75. Don't forget to attach the % symbol!	75%

Example. Change $\frac{5}{16}$ to a percent

Step 1.	Multiply the fraction by 100	$\frac{5}{16} \times 100$
Step 2.	Write 100 as a fraction	$\frac{5}{16} \times \frac{100}{1}$
Step 3.	Simplify before multiplication. 16 and 100 are both divisible by 4. Four goes into 100 twenty-five times; four goes into 16 four times. Multiply across top and bottom.	$\frac{5}{\underset{4}{\cancel{16}}} \times \frac{\overset{25}{\cancel{100}}}{1}$ $\frac{5 \times 25}{4 \times 1} = \frac{125}{4}$
Step 4.	Writing a percent in the form of an improper fraction defeats the purpose of being able to visualize the quantity in relation to 100. Change it to a mixed number*.	$\frac{125}{4} = 31\frac{1}{4}$
Step 5.	Add the % symbol. We're done.	$31\frac{1}{4}\%$

*Refer to 3.1.3 to change improper fractions to mixed numbers.

You-Try-It 2

Change to a percent:

a) $\frac{2}{5}$ b) $\frac{3}{8}$

Fractions that result in repeating decimals

See what happens when we try to write as a percent a fraction that, in its decimal form, repeats:

Example. Change $\frac{2}{3}$ to a percent

Step 1.	Multiply the fraction by 100; writing 100 as a fraction.	$\frac{2}{3} \times \frac{100}{1}$
Step 2.	No simplification before multiplication is possible. Multiply across top and bottom.	$\frac{2 \times 100}{3 \times 1}$ $\frac{200}{3}$
Step 3.	Writing a percent in the form of an improper fraction defeats the purpose of being able to visualize the quantity in relation to 100. Change it to a mixed number.	$\frac{200}{3} = 66\frac{2}{3}$
Step 4.	Don't forget to attach the % symbol.	$66\frac{2}{3}\%$

Fractions that correspond to repeating decimals will always have a fractional part when converted to a percent.

You-Try-It 3

Write as a percent:

a) $\frac{1}{6}$ 　　　　　　　　　　　　b) $\frac{5}{9}$

5.2.2.2 Changing an improper fraction to a percent

The process is no different than changing a proper fraction. Let's do one.

Example. Change $\frac{15}{6}$ to a percent

Step 1.	Multiply the fraction by 100	$\frac{15}{6} \times 100$
Step 2.	Write 100 as a fraction	$\frac{15}{6} \times \frac{100}{1}$
Step 3.	Simplify before multiplication. Six and 100 are both divisible by 2.	$\frac{15}{\underset{3}{6}} \times \frac{\overset{50}{\cancel{100}}}{1}$
Step 4.	Further simplification is possible before multiplying: Three is a factor of 15.	$\frac{\overset{5}{\cancel{15}}}{\underset{1}{\cancel{3}}} \times \frac{\overset{50}{\cancel{100}}}{1}$
Step 5.	Now, multiply across top and bottom. It results in a whole number: 250	$\frac{5 \times 50}{1 \times 1} = \frac{250}{1} = 250$
Step 6.	Add the % symbol. We're done.	250%

Note, in this result, 250% is the same as $2\frac{1}{2}$, which is the same as $\frac{5}{2}$.

The original fraction was $\frac{15}{6}$. We now see that if we had recognized at the start that $\frac{15}{6}$ reduces to $\frac{5}{2}$, the math would've been easier.

Lesson: Always make sure a fraction is reduced before doing anything with it. It can save you work in the long run.

You-Try-It 4

Change to a percent:

a) $\frac{17}{3}$

b) $\frac{16}{7}$

5.2.2.3 Changing a mixed number to a percent

When changing a mixed number to a percent there is a hard way, and a not-so-hard way. I'll show you both, but I recommend you use the not-so-hard way.

Method 1 (Harder)

Change the mixed number to an improper fraction and then follow the steps as shown in 5.2.2.2.

Example. Change $3\frac{2}{9}$ to a percent

Step 1.	Change the mixed number to an improper fraction	$3\frac{2}{9} = \frac{29}{9}$
Step 2.	Multiply the fraction by 100; writing 100 as a fraction.	$\frac{29}{9} \times \frac{100}{1}$
Step 3.	No simplification is possible; just multiply across top and bottom.	$\frac{29 \times 100}{9 \times 1} = \frac{2900}{9}$

Step 4.	Change the improper fraction to a mixed number.	$\dfrac{2900}{9} = 322\frac{2}{9}$
Step 5.	Add the % symbol. We're done.	$322\frac{2}{9}\%$

That wasn't so bad, but we ended up having to work with the fraction $\dfrac{2900}{9}$. Ugh! Luckily we can avoid it if we use the second method.

Method 2 (Not-so-hard)

Example. Change $3\frac{2}{9}$ to a percent

Before we get started on the example, a few words of explanation. A mixed number has a whole-number part and a (proper) fraction part. We remember how easy it was to change a whole number into a percent—you just add 2 zeroes. So, we will take advantage of this by splitting the mixed number up into its two parts, converting them separately into percents, and then just put them back together by adding the 2 percents.

Step 1.	Change the whole number, 3, into a percent by adding 2 zeros; then, file it away.	$3 = 300\%$
Step 2.	Now work with the (proper) fraction part: $\frac{2}{9}$. Multiply it by 100, written as a fraction.	$\dfrac{2}{9} \times \dfrac{100}{1}$
Step 3.	No simplification is possible; just multiply across top and bottom.	$\dfrac{2 \times 100}{9 \times 1} = \dfrac{200}{9}$
Step 4.	Change the improper fraction to a mixed number.	$\dfrac{200}{9} = 22\frac{2}{9}$
Step 5.	Add the percent amount from the whole number in Step 1, and add the % symbol.	$300 + 22\frac{2}{9} = 322\frac{2}{9}\%$

The advantage of this method over the first may not seem like such a big deal, but the whole number we were working with was only 3. For larger numbers, the improper fraction resulting from the mixed number conversion can produce daunting numbers that will be a pain to work with. Consider if you had to change $12\frac{2}{7}$ to a percent. After changing to an improper fraction and multiplying by 100, you would be looking at $\frac{8600}{7}$. Trust me on this one, method 2 is better.

You-Try-It **5**

Change to a percent:

a) $4\frac{4}{5}$

b) $17\frac{2}{3}$

5.2.3 Changing a decimal to a percent

Now that you've suffered through converting fractions, you get a break. Converting between decimals and percents is one of the easier things we do in math.

Multiplying a number by 10 just adds a zero: $6 \times 10 = 60$.

To multiply by 100, you just add 2 zeros: $6 \times 100 = 600$.

As we learned in *Multiplying decimals by powers of 10* in the previous chapter, to multiply a decimal number by 100 we just move the decimal point 2 places to the right.

To change a decimal number to a percent: move the decimal point 2 places to the right, then write the % symbol.

So, if we needed to change 0.6 to a percent, we just move the decimal point 2 places to the right, adding zeros if necessary, and then write the % symbol.   0.6 = 60%. Done.

It really is this easy.

Example. Change 0.13 to a percent

Step 1.	Move the decimal point 2 places to the right.	0.13 → 13
Step 2.	Write the % symbol.	13%

Example. Change 2.4 to a percent

Step 1.	Move the decimal point 2 places to the right. One zero needs to be added.	2.4 → 240
Step 2.	Write the % symbol.	240%

Example. Change 0.0625 to a percent

Step 1.	Move the decimal point 2 places to the right.	0.0625 → 6.25
Step 2.	Write the % symbol.	6.25%

This last example happens to be our sales tax rate around these parts...

You-Try-It 6

Change to a percent:

    a) 0.25   b) 2.435

Before we proceed to learning how to go in the reverse direction, changing percents to numbers, let's do the crazy fraction we started out talking about, $\frac{337}{1679}$. Because it produces such an (even more) unwieldy fraction, we will learn something new that will incorporate what we just learned about decimals. We will see that the most reasonable thing to do is to round the result, relying on the decimal-number value of the resulting fraction.

Example. Change $\frac{337}{1679}$ to a percent

Step 1.	Multiply the fraction by 100; writing 100 as a fraction.	$\frac{337}{1679} \times \frac{100}{1}$
Step 2.	No simplification before multiplication is possible. Multiply across top and bottom.	$\frac{337 \times 100}{1679 \times 1}$
Step 3.	The result is an unwieldy, ugly fraction.	$\frac{33700}{1679}$
Step 4.	Use a calculator to change to a decimal by doing the division, and round if necessary (rounded to the nearest whole percent).	$\frac{33700}{1679} \approx 20.07 \approx 20$

Step 5.	The value is very close to 20, so we give that as our result.	20%

You-Try-It answers

1a. 500% 1b. 3,000% 2a. 40% 2b. 375%

3a. $16\frac{2}{3}\%$ 3b. $55\frac{5}{9}\%$ 4a. $566\frac{2}{3}\%$ 4b. $228\frac{4}{7}\%$

5a. 480% 5b. $1,766\frac{2}{3}\%$ 6a. 25% 6b. 243.5%

Exercises 5.2

Write as a percent.

1. 0

2. 10

3. 7

4. 25

Convert to a percent.

5. $\frac{1}{4}$

6. $\frac{1}{5}$

7. $\frac{7}{10}$

8. $\frac{5}{8}$

9. $\frac{11}{16}$

10. $\frac{1}{3}$

11. $\frac{5}{6}$

12. $\frac{4}{9}$

Write as a percent.

13. $\dfrac{13}{3}$

14. $\dfrac{17}{5}$

15. $\dfrac{20}{9}$

16. $\dfrac{50}{4}$

Convert to a percent.

17. $4\dfrac{1}{2}$

18. $3\dfrac{2}{5}$

19. $2\dfrac{2}{3}$

20. $6\dfrac{1}{4}$

Write as a percent.

21. 0.75

22. 0.5

23. 0.125

24. 0.0625

25. 1.45

26. 10.09

27. Approximate $\dfrac{287}{1506}$ as a percent rounded to the nearest whole number

5.3 How to change a percent to a number

It's important to realize just *why* we change percents to actual numbers. We do it whenever we have to do a math computation with the percent! This will become abundantly clear in the upcoming sections where we do percent problems. If you're going to do some math with a percent, you must change the percent to an actual number.

When changing from percents to numbers we implement Rule 2: **Divide the percent by 100**. Makes sense. A percent is a number that has been blown up (multiplied) by a factor of 100. To undo that, to change it back, you have to undo multiplication by 100, which means you have to divide by 100.

With regards to *how* we do the division, what we do is deal with the percent on its terms. What I mean by this is that if the percent is in the form of a decimal, such as 4.75%, we should treat it as a decimal. If the percent is a whole number (like 35% or 217%), we can either treat it as a fraction or a decimal.

If the percent is given as a fraction or a mixed number, such as $\frac{3}{4}$% or $3\frac{5}{6}$%, then we should work with it as a fraction and look for producing a fraction as our answer. However, unless that is required, you may choose to convert fractions to decimals first, because the relationship between decimals and percents is so easy to work with – you just move the decimal point 2 places.

Let's start with some whole number percents, and see where it leads us...

5.3.1 Changing whole-number percents to decimals

To change a whole-number percent, such as 65%, you just put a decimal point after the number and then move it 2 places to the **left**—the direction that makes the number smaller. Remember, you are dividing a number by 100. You are making it smaller (a lot smaller). When you move a decimal point one place to the left, you are dividing the number by 10. For example

$$32 \div 10 = 3.2$$

If you move the decimal point two places to the left, you are dividing by 100—which is what we want to do. For example

$$32 \div 100 = 0.32$$

So, if we needed to change 32% to a number, we could give 0.32 as the answer. That would indeed be the simplest thing to do.

The alternative would be to write it as a fraction.

Let's try some.

Example. Write 450% as a decimal number

Step 1.	Write the number with a decimal point.	450.
Step 2.	Move the decimal point 2 places to the left.	4.50
Step 3.	Remove any trailing zeros in the decimal part.	4.5

Example. Write 64% as a decimal number

Step 1.	Write the number with a decimal point.	64.
Step 2.	Move the decimal point 2 places to the left.	.64
Step 3.	Since there is no whole-number part, write a leading zero in the ones place.	0.64

You-Try-It 1

Write the percent as a decimal:

a) 84%   b) 1,085%

5.3.2 Changing whole-number percents to fractions

Now let's convert percents to fractions, either as proper fractions or as mixed numbers, as the case may be.

The alternative to writing 32% as the decimal number 0.32 is to turn it into a fraction. What you need to do is divide 32 by 100: $32 \div 100$. Fractions mean division. A " $\div$ " symbol is no different than a fraction bar.

$$32 \div 100 = \frac{32}{100}$$

Then, you reduce.

$$\frac{32 \div 4}{100 \div 4} = \frac{8}{25}$$

So, $32\% = \frac{8}{25}$.

Example. Change 88% to a fraction

Step 1.	Write the number in a fraction with denominator 100 .	$\frac{88}{100}$
Step 2.	Reduce/simplify, if possible. Four is a common factor of both 88 and 100. The fraction reduces.	$\frac{88 \div 4}{100 \div 4} = \frac{22}{25}$

Example. Change 415% to a fraction

Step 1.	Write the number in a fraction with denominator 100.	$\frac{415}{100}$
Step 2.	The fraction is improper. We could leave it improper and just reduce it, or change it to a mixed number. In both cases it will need to be reduced, so we will reduce first. Five is a common factor of 415 and 100.	$\frac{415 \div 5}{100 \div 5} = \frac{83}{20}$
Step 3.	If we don't want to leave it improper, we can change it to a mixed number.	$\frac{83}{20} = 4\frac{3}{20}$

You-Try-It 2

Write the percent as a fraction. If improper, write also as a mixed number.

    a) 78%   b) 636%

5.3.3 Changing decimal-number percents to decimal numbers

If you have a percent that has a decimal number part such as 5.75% or 0.75% that needs to be used in a computation, the simplest thing is to change it to its decimal number equivalent. As we showed in 5.3.1, we just have to move the decimal point two places to the left, which is how we divide a decimal number by 100.

Example. Change 5.75% to a decimal number

Step 1.	Write the number without the "%".		5.75
Step 2.	Move the decimal point 2 places to the left. Fill with zeros if necessary.		.0575
Step 3.	If there is no whole-number part, write a 0 in the ones place.		0.0575

Example. Change 0.25% to a decimal number

Step 1.	Write the number without the "%".	0.25
Step 2.	Move the decimal point 2 places to the left. Fill with zeros if necessary.	.0025
Step 3.	If there is no whole-number part, write a 0 in the ones place.	0.0025

Example. Change 275.5% to a decimal number

Step 1.	Write the number without the "%".	275.5
Step 2.	Move the decimal point 2 places to the left.	2.755

You-Try-It 3

Write the percent as a decimal:

a) 83.4%

b) 108.5%

c) 0.00675%

d) 2.399%

5.3.4 Changing fractional and mixed-number percents to numbers

The most difficult percent conversions are—*surprise!*—those involving fractions and mixed numbers. If you have a percent that you need to use in a computation, such as an interest rate like $2\frac{3}{4}$ %, we have two choices. We can either convert it into a decimal or a fraction. If we know beforehand that we want to have a decimal number to use in our computation, then the easiest route is to first convert $2\frac{3}{4}$ to a decimal number and then just move the decimal point.

If for some reason we prefer to have a fraction to use in our computation, then we will not convert it to a decimal first, but simply do the fraction division, i.e., divide $2\frac{3}{4}$ by 100 using fraction division.

We'll start by doing each of these in a step-by-step example.

Example. Change $2\frac{3}{4}$ % to a decimal number (Decimal method)

Step 1.	Change $2\frac{3}{4}$ to a decimal. We recognize $2\frac{3}{4}$ as $2+\frac{3}{4}$. We recognize that we just need to find the decimal equivalent of $\frac{3}{4}$, and then add it to 2.	$\frac{3}{4} \rightarrow 3 \div 4$ $3 \div 4 = 0.75$
Step 2.	Add the decimal number to 2.	$2 + 0.75 = 2.75$
Step 3.	Divide by 100 → move the decimal point 2 places to the left.	0.0275

Example. Change $2\frac{3}{4}$ % to a fraction (Fraction method)

Step 1.	Since we have to do fraction division, we have to first change $2\frac{3}{4}$ to an improper fraction.	$2\frac{3}{4} = \frac{?}{4}$

Step 2.	We get the numerator to the improper fraction by multiplying the denominator, 4, by the whole-number part, 2, and adding the numerator, 3.	$4 \times 2 + 3 = 11$ $2\dfrac{3}{4} = \dfrac{11}{4}$
Step 3.	Divide the improper fraction by by 100. The result is $\dfrac{11}{400}$.	$\dfrac{11}{4} \div 100$ $\dfrac{11}{4} \div \dfrac{100}{1}$ $\dfrac{11}{4} \times \dfrac{1}{100}$ $\dfrac{11 \times 1}{4 \times 100} = \dfrac{11}{400}$

You should not be surprised at getting unusual fractions after converting from small percents. This is typical, as the next example shows.

Example. Change $12\dfrac{5}{8}\%$ to a fraction

Step 1.	Since we have to do fraction division, we have to first change $12\dfrac{5}{8}$ to an improper fraction.	$12\dfrac{5}{8} = \dfrac{?}{8}$
Step 2.	We get the numerator to the improper fraction by multiplying the denominator, 8, by the whole-number part, 12, and adding the numerator, 5.	$8 \times 12 + 5 = 101$ $12\dfrac{5}{8} = \dfrac{101}{8}$

Step 3.	Divide the improper fraction by by 100. The result is $\frac{101}{800}$.	$\frac{101}{8} \div 100$ $\frac{101}{8} \div \frac{100}{1}$ $\frac{101}{8} \times \frac{1}{100}$ $\frac{101 \times 1}{8 \times 100} = \frac{101}{800}$

You-Try-It 4

Write the percent as a decimal number:

a) $6\frac{1}{4}\%$

b) $75\frac{1}{2}\%$

c) $\frac{3}{5}\%$

d) $5\frac{1}{8}\%$

Write the percent as a fraction:

    e) $6\frac{1}{4}\%$ f) $75\frac{1}{2}\%$

    g) $\frac{3}{5}\%$ h) $5\frac{1}{8}\%$

Why use fractions when decimals are usually so much easier to work with?

There is one situation where it is preferable to change a fraction/mixed-number percent to a fraction rather than to a decimal: when the decimal representation of the fraction is a repeating decimal, like $\frac{1}{3}$ or $\frac{1}{6}$.

The following example, done both ways, shows why.

Example. Change $5\frac{2}{3}\%$ to a fraction

Step 1.	Since we have to do fraction division, we have to first change $5\frac{2}{3}$ to an improper fraction.	$5\frac{2}{3} = \frac{?}{3}$
Step 2.	We get the numerator to the improper fraction by multiplying the denominator, 3, by the whole-number part, 5, and adding the numerator, 2.	$3 \times 5 + 2 = 17$ $5\frac{2}{3} = \frac{17}{3}$
Step 3.	Divide the improper fraction by by 100. The result is $\frac{17}{300}$.	$\frac{17}{3} \div 100$ $\frac{17}{3} \div \frac{100}{1}$ $\frac{17}{3} \times \frac{1}{100}$ $\frac{17 \times 1}{3 \times 100} = \frac{17}{300}$

Let's see why this is preferable to changing it to a decimal:

Example. Change $5\frac{2}{3}\%$ to a decimal number

| Step 1. | First change $5\frac{2}{3}$ to a decimal.

 We recognize $5\frac{2}{3}$ as $5 + \frac{2}{3}$.

 We recognize that we just need to find the decimal equivalent of $\frac{2}{3}$, and then add it to 5. | $\frac{2}{3} \rightarrow 2 \div 3$

 $2 \div 3 = 0.\overline{6}$ |

Step 2.	$\frac{2}{3}$ is a repeating decimal: 0.6666... How many 6s do we include? A judgment call is necessary...	$\frac{2}{3} = 0.6666\ldots$
Step 3.	Add the decimal number to 5.	$5 + 0.6666\ldots = 5.6666\ldots$
Step 4.	Divide by 100 → move the decimal point 2 places to the left.	$0.056666\ldots$
Step 5.	What do we write as our final answer? There are many options... All but the last are considered approximations.	$5\frac{2}{3}\% \cong 0.056666$ $5\frac{2}{3}\% \cong 0.0567$ $5\frac{2}{3}\% \cong 0.057$ $5\frac{2}{3}\% = 0.05\overline{6}$

The fractional representation of $5\frac{2}{3}\%$ is superior to the decimal representations because <u>*all of the numerical information is contained within the simple fraction,*</u> $\frac{17}{300}$. One whole number, 17, divided by another, 300. There is no ambiguity. No judgment call to be made when using it in a calculation. No bogus writing a bar over a digit to represent an infinite number of digits beyond the decimal point.

You-Try-It 5

Write the percent as a decimal number:

a) $6\frac{1}{3}\%$ b) $4\frac{4}{9}\%$

Write the percent as a fraction:

c) $6\frac{1}{3}\%$

d) $4\frac{4}{9}\%$

You-Try-It answers

1a. 0.84

1b. 10.85

2a. $\frac{39}{50}$

2b. $6\frac{9}{25}$

3a. 0.834

3b. 1.085

3c. 0.0000675

3d. 0.02399

4a. 0.0625

4b. 0.755

4c. 0.006

4d. 0.05125

4e. $\frac{1}{16}$

4f. $\frac{151}{200}$

4g. $\frac{3}{500}$

4h. $\frac{41}{800}$

5a. $0.06\overline{3}$

5b. $0.0\overline{4}$

5c. $\frac{19}{300}$

5d. $\frac{2}{45}$

Exercises 5.3

Write as a decimal or whole number.

1. 56%

2. 5%

3. 140%

4. 2,000%

Write as a fraction (simplify/reduce as necessary).

5. 75%

6. 89%

7. 20%

8. 6%

Write as an improper fraction and also as a mixed number.

9. 320%

10. 125%

11. 260%

12. 448%

Write as a decimal number.

13. 72.2%

14. 15.5%

15. 112.8%

16. 0.025%

17. 0.12%

18. 3.99%

Write as a decimal number.

19. $5\frac{1}{2}\%$

20. $\frac{3}{4}\%$

21. $65\frac{4}{5}\%$

22. $6\frac{3}{8}\%$

Write as a fraction.

23. $5\frac{1}{2}\%$

24. $\frac{3}{4}\%$

25. $65\frac{4}{5}\%$

26. $6\frac{3}{8}\%$

Write as a decimal number.

27. $4\frac{2}{3}\%$

28. $5\frac{1}{9}\%$

Write as a fraction.

29. $4\frac{2}{3}\%$

30. $5\frac{1}{9}\%$

5.4 Solving basic percent problems

The most common place we see percents in daily life is in financial transactions. A department store may have an item on sale for 20% off. Let's say you are looking at a jacket that is marked down by 20%. If the regular price was $50, you need to know what the discount is, and then calculate the sale price of the jacket.

5.4.1 The three basic percent quantities and their corresponding problem types

There are always three basic quantities involved in a percent situation like this one. In the sale-price jacket example the quantities are:

- The regular price of the jacket; in this case $50
- The percent discount; in this case 20%
- The discount amount; in this case what needs to be calculated, namely, 20% of $50, (which is $10).

In any given percent problem two of these quantities will be known and you will need to find the third one, the unknown quantity.

Since calculating a sale price from a percent discount is only one of many different situations that can lend themselves to a percent problem, we provide generic names for each of these three quantities.

In our example, if we strip out the money units from the numbers, the essential mathematics is this:

$$20\% \text{ of } 50 \text{ is } 10.$$

The '20%' quantity is called—surprise, surprise—the **Percent**.

The '10' is often called the "Amount." I prefer to call it the "**Part**." It is, after all, the *part* of 50 corresponding to 20% of it.

The '50' is often called the "Base." I prefer to call it the "**Whole**," as in, when we take 20% of *it* we get 10. So the '*it*'—the $50 in the example—is the *whole* amount, of which the 20% is the *part*.

Just to summarize my vocabulary for those who are more familiar with the words often used elsewhere:

$$\text{Part} = \text{Amount}$$
$$\text{Whole} = \text{Base}$$

You can use these interchangeably if you like.

Learning how to solve basic percent problems requires learning how to find each of the three quantities in a basic percent problem. We have to learn which calculations to do for each of the three different types of problems. Before we begin by learning how to find the Part, one thing needs mentioning.

In general there are many different types of math problems that involve percents. At some level, they all rely on solving one of these three *basic* types of percent problems, but the actual question that needs to be answered may be something different. A good example of this is the one we have just used. In the real world the quantity that you're really concerned with is not the discount amount of a $50 jacket on sale at 20% off, but what you're actually going to be charged for it at the register (before tax), in other words, the sale price.

It doesn't take a rocket scientist to recognize that all you need to do is subtract the $10 discount amount **(Part)** from the $50 regular price **(Whole)**, arriving at $40. So, one extra calculation is required beyond the *basic* calculation of finding the Part to actually answer the critical question: "How much is this going to cost me?"

Another common type of percent problem is finding a *Percent Increase* or *Decrease*. Here the final calculation is finding the Percent, but extra calculations have to be done *beforehand* to determine the Part. We will learn how to do these problems in 5.6, *Common percent problems*.

5.4.2 The "Triangle" method for solving basic percent problems

There are many different methods for solving percent problems. None should be thought of as *better* than another. It's really a matter of personal preference. When I teach percents in an algebra class I usually show three different methods (two of which require algebra). I encourage the students to learn how to do all of them when doing the homework assignments, but then to use whichever they prefer when taking the test. *Because everyone is different.* Repeat after me: *Everyone is special.* It's a lifestyle choice!

Some students will prefer one of the algebraic methods exclusively, another may gravitate toward the non-algebraic approach, and still others will use one method for one type and prefer a different method for another type of problem. Because this book presumes you have no knowledge of algebra, I will first show you a visual, non-algebraic method called the **Triangle method**, and then in the next section show another method that relies on formulas.

The Triangle method requires filling in numbers in two of the sections (boxes) of a triangle that is divided into three parts, and then remembering what operation is required on the two numbers. Here is the Triangle with the sections labeled:

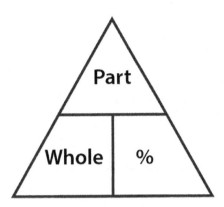

Here is the triangle with the code words substituted in the sections so it is even easier to fill in the numbers:

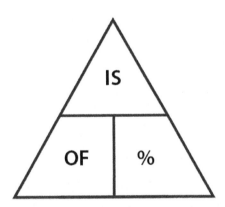

When doing a percent problem using this method there will be numbers in two of the boxes. In the remaining box you can write a "?" if you like, to represent the unknown quantity, or just leave it blank. Here are the two rules for doing the calculations:

1. If the numbers are in the bottom two sections, you multiply them to get the answer (what belongs in the top section, where the "?" is). In this case, you are finding the Part.
2. If one number is in the top section, you divide it by the number that is in either of the bottom sections. (If the number is in the bottom-left "Whole/OF" section, you are finding the Percent. If the number is in the bottom-right "%" section, you are finding the Whole.)

This is illustrated in the next diagram of the triangle, *which contains all the information you need to solve any of the three basic percent problem types*, with the proviso that if a percent is entered it must be converted to a number first:

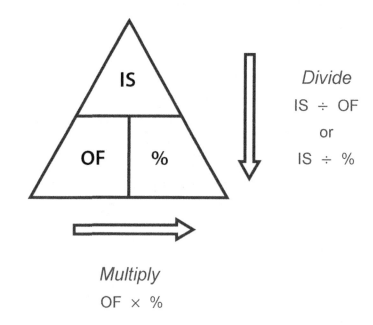

Divide

IS ÷ OF

or

IS ÷ %

Multiply

OF × %

Now we are ready to do some examples of each type.

5.4.3 Solving percent problems when the Part is unknown

We might as well start by using our original example: What is 20% of $50? Notice the words "is" and "of" in the sentence.   The Percent is known: 20%. One thing we must remember is that when the percent is entered into the triangle, it must be as a number! Refer to 5.3 if you need to, to change 20% to either the *decimal number* 0.2 or the *fraction* $\frac{1}{5}$.

Using the decimal version 0.2, we put it into the "%" section at the bottom right. "$50" is attached to "of." We put 50 into the other bottom section of the triangle. The word "what" is attached to "is," indicating that the unknown quantity goes in the top box, so our triangle looks like this:

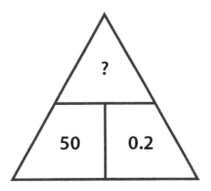

Since both of the bottom sections of the triangle are filled in, we have to multiply:

$$50 \times 0.2 = 10$$

The answer is 10, and to the problem at hand, $10. The discount on the jacket amounts to $10. So, you will be charged $40 ($50 – $10) for it at the register.

Example. In the Bizarro world, *junior* citizens get a 15% discount at Denny's. What is the discount on a "Moons over my-hammy" breakfast normally costing $7.99? What is the actual discounted price?

This question boils down to: What is 15% of 7.99? ... and then subtracting it from $7.99.

Step 1.	Fill in the boxes in the triangle. We put 7.99 in the "OF" box and 0.15 in the "%" box (because 15% as a decimal number is 0.15). We have put a '?' in the top box, which indicates that the Part is the unknown quantity.	 ? 7.99 0.15
Step 2.	Do the calculation. Since we have numbers in the bottom two boxes, we have to multiply them to get the Part:	$0.15 \times 7.99 = 1.1985$ This rounds to $1.20. This is the discount.
Step 3.	To find the discounted price we subtract the $1.20 discount from the regular price.	$7.99 - 1.20 = 6.79$ The discounted price is $6.79.

You-Try-It **1**

    a) What is 40% of 1,025?

    b) The property tax rate where you live is 2%. What is the property tax on a house valued at $165,000?

    c) What is 140% of 65?

    d) The bill for your meal is $18.30. You wish to tip 20%. What is the total amount you will pay (excluding tax)?

5.4.4 Solving percent problems when the Whole is unknown

One of the reasons that the word "Base" is often used rather than "Whole" is because in percent problems the Part can be bigger than the Whole. For instance, when something doubles in size it has grown by 100%. If something grows by one-and-a-half its original size, it has grown by 150%. (These are examples of *percent increase* that we tackle in 5.6, *Common percent problems*.)

We keep this in mind as we tackle some percent problems where the Whole is the unknown quantity. The Whole will turn out to be less than the Part if the percent is greater than 100.

Example. Seventy-five is 80% of what number?

Step 1.	Fill in the boxes in the triangle. Seventy-five belongs to "is" and so goes in the top box. We put 0.8 in the percent box at the bottom-right (because 80% as a decimal number is 0.8). We have put a '?' in the OF box. (Notice in the sentence the words "what number?" is associated with "of", implying the unknown quantity.)	
Step 2.	Do the calculation. Since we have numbers in the top and bottom, we have to divide 75 by 0.8 to get the Whole.	$75 \div 0.8 = 93.75$ Our answer: 75 is 80% of <u>93.75</u>

Example. The National Debt has grown to 225% of what it was 10 years ago. If it is now $18trillion, what was it in 2005?

(We do not need to work with the number 18 trillion in its standard form, i.e., 18,000,000,000,000. Instead, we will work with the much simpler number 18, and just remember that any numbers we produce from working with it will also be in *trillions*. When done we can write the resulting number, if we choose, in standard form.)

This question boils down to: 18 is 225% of what number?

Step 1.	Fill in the boxes in the triangle. Eighteen belongs to "is" and so goes in the top box. We put 2.25 in the percent box at the bottom-right (because 225% as a decimal number is 2.25). We have put a '?' in the OF box. (Notice in the sentence the words "what number?" is associated with "of," implying the Whole is the unknown quantity.)	

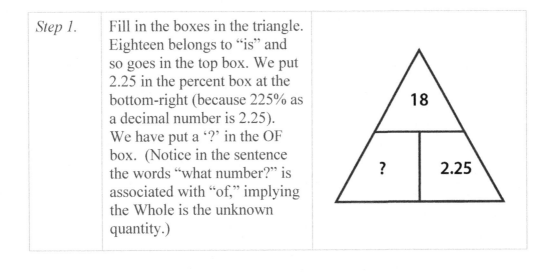

Step 2.	Do the calculation. Since we have numbers in the top and bottom, we have to divide 18 by 2.25 to get the Whole.	$18 \div 2.25 = 8$ The National Debt was $8 trillion in 2005, or $8,000,000,000,000 in standard notation.

You-Try-It 2

a) Twenty-five is 25% of what number?

b) Forty-eight is 12% of what number?

c) 189,000 is 140% of what number?

d) $5\frac{1}{2}$ is 11% of what number?

5.4.5 Solving percent problems for the Percent

When solving for the Percent an extra bit of work is required. Depending on the method we use we may need to change the number to a percent after doing the calculation. *This holds for the Triangle method.* So, once doing the calculation, you will have to multiply the answer by 100 to turn it into a percent. Refer to 5.2 for a reminder on how this works.

Example. 208 is what percent of 260?

Step 1.	Fill in the boxes in the triangle. 208 goes into the "IS" box on top. 260 goes into the "OF" box, bottom-left. We can put a '?' in the "%" box.	 208 260 | ?
Step 2.	Do the calculation. Since we have numbers in the top and bottom, we have to divide 208 by 260 to get the percent value.	$208 \div 260 = 0.8$ The answer, 0.8, is not yet a percent.
Step 3.	Change the percent number to an actual percent by multiplying by 100. *Just move the decimal point 2 places to the right.*	$0.8 \rightarrow 80.0$ Our answer is 80%.

You-Try-It 3

a) Fifty is what percent of 200?   b) Twenty-five is what percent of 400?

c) $97.50 is what percent of $780?   d) 110 is what percent of 40?

You-Try-It answers

1a. 410	1b. $3,300	1c. 91	1d. $21.96
2a. 100	2b. 400	2c. 135,000	2d. 50
3a. 25%	3b. 6.25%	3c. 12.5%	3d. 275%

Exercises 5.4

Use the triangle method to solve.

1. Find 25% of 148.

2. What is 16% of 250?

3. Find 78% of 50.

4. What is 120% of 55?

5. Sixty-two is 40% of what number?

6. 3,500 is 125% of what number?

7. 12% of what number is $10\frac{1}{5}$?

8. 25 is $6\frac{1}{4}$% of what number?

9. 72 is what percent of 240?

10. Thirty-five is what % of 500?

11. 145 is what % of 8?

12. 2,400 is what percent of 150,000?

5.5 Using formulas to solve basic percent problems

This purpose of this book is to teach elementary math. It does not cover signed (negative) numbers, which would be the next topic if you were on the path to learn all of what is considered "pre-algebra" math. Also it does not attempt to teach algebra. However, I recognize that some readers will be interested in learning more than the Triangle method to solve problems involving percentages. It is practically impossible to go beyond the Triangle method, or some similar visual method, without encountering algebra in one form or another.

This section is for those readers who wish to learn more than just a "trick" method for solving percentage problems. *Other readers can consider it optional.* I will use algebra in the most minimal sense possible. In fact, it's possible to go through this section without even realizing that you are using algebra.

Broadly speaking, when we think of algebra there are two things that usually come to mind first:

1. The notion of representing an unknown number with the letter x (or any letter). This is representation of an unknown quantity by a *variable*.
2. The idea of solving an *equation*. This is a mathematical expression that includes an equal (" = ") sign.

With regards to (1), I will use the letter x when representing one of the unknown quantities in a basic percent problem (namely: the Part, Whole, or Percent) or simply use the words that name them.

With regards to (2), some equations can be thought of as *formulas*. For instance, the formula for finding the Part in a basic percent problem is:

$$Part = Percent \text{ x } Whole$$

You can think of this equation as a formula for finding the Part when you know the Percent and the Whole. If it helps, you can also think of this as a *recipe*—yes, just like a recipe for baking a cake.

A recipe for cake could be thought of mathematically as a formula to find *Cake*:

$$Cake = \{so\ much\}\ Flour + \{so\ much\}\ Filling\ Ingredients + \{so\ much\}\ Egg,\ etc...$$

(I'm not a baker; I buy my cakes...)

In this formula, "so much" would be replaced with a specific quantity, as in say, *2*, for "so much" Egg, or *1 cup*, for "so much" flour.

Similarly a formula for finding the Part in a percent problem could be considered a *recipe* for finding it *given the quantities Percent and Whole.*

I hope this helps. We will need no more knowledge of algebra than this to work through solving the basic percent problems using formulas.

5.5.1 Finding the Part

As revealed above, the formula for finding the Part in a basic percent problem is:

$$Part = Percent \times Whole$$

A very important thing to bear in mind as we learn to use algebraic formulas to solve percent problems is that *the Percent must be given as a number.* For example, 50% must be represented either in its decimal form 0.5, or as the fraction ½. This holds for all the problems we will be doing in this section.

When we need to find the Part, the formula tells us that all we have to do is multiply the Percent by the Whole. Here is what is known as a direct translation from words to algebra:

What   is   20% of 75?

$$Part   =   0.2 \times 75$$

For those interested, in standard algebra, *Part,* the unknown quantity, would be represented by a variable, typically x, so the formula would actually look like this:

$$x = 0.2 \cdot 75$$

Notice, for obvious reasons, we stop using the cross symbol (which is no different than the letter 'x') to show multiplication. We can either use the dot, "·", as I did here, or use parentheses to indicate the multiplication, like this:

$$x = (0.2)(75)$$

Feel free to use any of these variations in solving the problems in this section.

Example. A hardware store has a 15%-OFF sale on all space heaters. You want to buy one that is regularly priced at $45. What is the discount amount, and what is the sale price?

This question boils down to finding 15% of 45 and then subtracting that amount from $45.

Step 1.	We first recognize that we are looking for the *Part* in a basic percent problem. We remember the formula for finding the Part is: Part = Percent **x** Whole. We know the Percent must be changed to a number. We choose to use its decimal representation of 0.15.	Part = Percent **x** Whole *Part* = 15% **x** 45 *Part* = 0.15 **x** 45
Step 2.	Do the calculation.	$0.15 \times 45 = 6.75$ $6.75 is the discount amount.
Step 3.	To find the sale price we subtract the $6.75 discount from the regular price.	$45 - 6.75 = 38.25$ The sale price is $38.25.

You-Try-It 1

a) What is 65% of $895?　　　　b) What is 220% of 65?

c) The total hotel tax rate in Boston is 14.45%. What is the total daily rate for a room priced at $140 a night?

d) The bill for your hotel meal came to $58.60. You were dissatisfied with the food *and the service* and wish to tip only 12%. What is the total amount you will pay (excluding tax)?

5.5.2 Finding the Whole

The formula for finding the Whole in a basic percent problem is:

$$Whole = Part \div Percent$$

(This is evident from the Triangle method, where we divide the number in the top, *Part*/IS box of the triangle, by the *Percent* in the bottom box, to get the *Whole*. Refer to the diagram in 5.4.2)

As you might expect, just as we don't use the standard cross for multiplication in algebra, we also don't use the division symbol (" $\div$ "). Since division is really the same thing as a fraction (top divided *by* bottom), we write division as a fraction. Our formula for finding the Whole when we know the Part and the Percent looks like this:

$$Whole = Part \ / Percent$$

or

$$Whole = \frac{Part}{Percent}$$

And just as before, the Percent must be written as a number—in its whole, decimal, or fraction equivalent.

Example. If the tax on an auto repair was $38.25 and the sales tax rate is 6.25%, what was the charge for the repair?

This question boils down to asking: 6.25% of what number is 38.25? We first recognize that the unknown quantity that we are looking for is the Whole, as the words "what number" follows the word "of".

After turning 6.25% into its decimal equivalent 0.0625 (refer to 4.3.2), direct algebraic translation leads to this:

$$6.25\% \text{ of what} \quad \text{is} \quad 38.25?$$

$$0.0625 \text{ of Whole} \quad = \quad 38.25$$

Notice how in this case the direct translation does *not* give us a formula per se—in that we do not have the unknown quantity (Whole) by itself on one side of the equation. Part of learning algebra is learning how to change the equation

$$0.0625 \text{ of } x = 38.25$$

into the equation

$$x = 38.25/0.0625$$

But we don't have to know how to do that because we already know the formula for finding the Whole when we know the Part and the Percent; it is

Whole = Part / Percent.

We can just fill in the two quantities and do the math—after, of course, turning 6.25% into its decimal representation, 0.0625:

$$\text{Whole} = 38.25 / 0.0625$$

Which turns out to be 612. So the charge for the repair work was $612.

Example. 455 is 125% of what number?

Step 1.	We recognize the unknown quantity is the Whole because the words "what number?" follows the word "of." We use the formula for finding the Whole: Whole = Part/Percent writing 125% as the decimal 1.25.	Whole = Part / Percent *Whole* = 455 / 1.25

Step 2.	Do the calculation. Divide 455 by 1.25.	$455 \div 1.25 = 364$ Our answer: 364.
Step 3.	Confirm our result. Is 455 equal to 125% of 364? Yes.	? $1.25 \times 364 = 455$ $455 = 455$

You-Try-It 2

    a) Forty-five is 90% of what number?   b) 5.07 is 13% of what number?

    c) 209,250 is 135% of what number?   d) $15\frac{2}{5}$ is 7% of what number?

5.5.3 Finding the Percent

The formula for finding the Percent in a basic percent problem is:

$$Percent = Part \div Whole$$

(This is evident from the Triangle method, where we divide the number in the *Part*/IS box of the triangle by the *Whole*/OF in the bottom box. Refer to the diagram in 5.4.2)

There is one important difference between using this formula and using the formulas to find the Part or the Whole. Recall from the Triangle method, when the Percent was the unknown we had to do an additional step after doing the calculation: change the number for the Percent into an actual percent, i.e., multiply it by 100.

We have to do the same thing when we use the formula. If we look at this formula in its purely algebraic form,

$$Percent = \frac{Part}{Whole}$$

we see that the right side of the formula is actually a fraction. But, after doing the division, it isn't a percent until it is multiplied by 100. Why not incorporate this calculation into the formula itself so we have only one thing to memorize, and not run the risk of forgetting to do it?

Here, then, is the formula we will use for finding the Percent when the Part and the Whole are known:

$$Percent = \frac{Part}{Whole} \times 100\%$$

And just to be extra nice we include the " % " symbol in the formula to make sure we don't forget to write it.

Example. Sixty-four is what percent of 80?

We recognize that the unknown quantity we are looking for here is the percent. Let's go right to formula, and plug the numbers in:

$$Percent = \frac{64}{80} \times 100\%$$

As far as doing the calculation goes, we have a division and a multiplication to do. We can do this in any order, either:

$$64 \div 80 \times 100 \text{ or}$$

$$64 \times 100 \div 80$$

(or even $100 \div 80 \times 64$ or $100 \times 64 \div 80$).

The answer is 80%. ***Do NOT forget to write the " % " symbol!***

Example. Forty-five is what percent of 12?

Step 1.	We recognize that we are looking for the Percent. The Part is 45, and the Whole is 12 (note 45 precedes "is" and 12 follows the word "of"). Plug the two numbers into the formula for finding the Percent.	$Percent = \dfrac{Part}{Whole} \times 100\%$ $Percent = \dfrac{45}{12} \times 100\%$
Step 2.	Do the calculation. $45 \div 12 \times 100$	$45 \div 12 \times 100 = 375$ The answer is 375%.

Note, the percent is greater than 100% because in this problem the Part is greater than the Whole.

You-Try-It 3

a) Eighty is what percent of 250?   b) 2.5 is what percent of 500?

c) $29.25 is what percent of $450?   d) 180 is what percent of 50?

You-Try-It answers

1a. $581.75   1b. 143   1c. $160.23   1d. $65.63

2a. 50   2b. 39   2c. 155,000   2d. 220

3a. 32%   3b. 0.5%   3c. 6.5%   3d. 360%

Exercises 5.5

Use formulas to solve.

1. Find 180% of 45.

2. What is 35% of $775?

3. Find 6.25% of 4,800.

4. What is $3\frac{1}{5}$% of 125?

5. Sixty-four is 80% of what number?

6. 27 is 15% of what number?

7. 13.6 is 85% of what number?

8. $45\frac{1}{2}$ is 130% of what number?

9. $22\frac{3}{4}$ is what percent of 650?

10. 2.2 is what % of 8?

11. 114 is what % of 5?

12. Seven is what percent of 20?

5.6 Common percent problems

Before we leave percents, let's tackle some typical real-world problems that use them. I've already snuck a couple into the preceding sections, so let's add just a few more to give a more complete picture.

5.6.1 Tax

When a quantity is taxed at a certain *tax rate* the result is the *tax amount*, which we normally just call the *tax*. For example, if the sales tax is 6% and you buy a TV for $500, the *tax* is $30.

The math involved looks like this:

$$Tax = tax\ rate \times cost$$

$$\$30 = 6\% \times \$500$$

$$30 = 0.06 \times 500$$

This formula, *Tax = tax rate x cost* , is really no different than the basic formula for finding the Part: *Part = Percent x Whole* .

Recognizing this we realize that any type of tax problem can be done using either the Triangle method or the algebraic formula methods from 5.5, providing we make the following substitutions:

$$Tax = Part$$

$$Cost = Whole$$

$$Tax\ rate = Percent.$$

Example. A city has a property tax rate of 1.7%. This year they have assessed the value of your home at $145,000. What is your property tax?

$$Tax = Part = the\ unknown$$

$$Cost = Whole = \$145,000$$

$$Tax\ rate = Percent = 1.7\% => 0.017$$

To find the Part in a basic percent problem using the Triangle method:

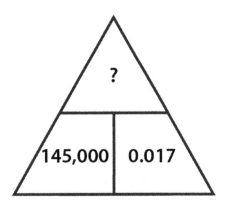

The Triangle method requires that we multiply when numbers appear in the bottom two boxes; so, the property tax is arrived at by multiplying 145,000 by 0.017. The property tax is $2,465.

Or, we could use the formula from 5.5.1,

$$Part = Percent \text{ x } Whole$$

$$Tax = 0.017 \text{ x } 145,000$$

$$Tax = \$2,465.$$

Example. You paid a sales tax of $37.50 on a video camera that cost $600. What was the sales tax rate?

Here the Percent is the unknown quantity; we are given the values for the Part (tax) and the Whole (cost).

$$Tax = Part = \$37.50$$

$$Cost = Whole = \$600$$

$$Tax\ rate = Percent = the\ unknown$$

The formula from 5.5.3 for finding the Percent is

$$Percent = \frac{Part}{Whole} \times 100\%$$

Plugging 37.5 into the Part and 600 into the Whole in the formula gives us:

$$Percent = \frac{37.5}{600} \times 100\%$$

And doing the calculation gives us:

$$Percent = Sales\ tax\ rate = 6.25\%$$

If we prefer to use the Triangle method, it looks like this:

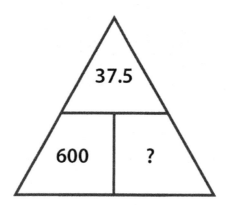

Because a number appears in the top box, we divide its value, 37.5, by the number underneath it, 600:

$$Percent = 37.5 \div 600.$$

The result of the division is 0.0625, and we must remember to turn this number into a percent. We multiply it by 100, which is done by moving the decimal point 2 places to the right:

$$Percent = 6.25\%.$$

This is the sales tax rate.

You-Try-It 1

a) A new federal "Internet Tax" has been imposed. It requires any online sales to pay a tax of 1.2%. Your online business generated $2,355 in a given month. What is the amount of Internet Tax that you owe?

b) You paid an import tax of $7,500 on a Ferrari you purchased from Italy for $125,000. What was the import tax rate?

c) The property tax rate in your city is 1.75%. If your property tax for the year is $2,800, what is the assessed value of your home?

5.6.2 Commission

Many jobs pay on a commission basis. The pay may be totally based on commission or there may be a base salary on top of which a commission component is added. The commission amount is usually calculated as some percent of the total sales made by the employee during some time period.

Example. A jewelry sales position pays a base salary of $1,200 per month + 3% of all sales for the month. If a salesperson sold $18,025 in a given month, what is her pay for the month?

This problem boils down to calculating the commission she made on the $18,025 total sales and then adding it to the base salary:

Monthly pay = 1200 + 3% of 18,025.

So, we need to find 3% of 18,025. Once again, we are seeking the Part, given the Whole and the Percent.

Let's use the formula: *Part = Percent* x *Whole*, recognizing that the commission is the Part, the total sales amount is the Whole, and commission rate is the Percent.

commission = 3% of $18,025

commission = 0.03 x 18,025

commission = $540.75.

The commission of $540.75 has to be added to the base salary to arrive at the total monthly pay:

Monthly pay = base + commission

Monthly pay = 1200 + 540.75 => $1,740.75

Example. A real estate salesman's income is based purely on commission. He is paid a 1.2% commission on his sales.   If his pay for the month of March was $15,084, what were his total sales?

Let's pair up the quantities in a commission problem with the 3 basic percent quantities.

Commission= Part

Commission rate = Percent

Total sales = Whole

Now let's fill in the quantities from the problem:

Commission = Part = 15,084

Commission rate = Percent = 1.2% => 0.012

Total sales = Whole = the unknown

It's now evident that this is a basic percent problem where the Part and Percent are known and the Whole is the unknown quantity, which we need to find.

The formula from 5.5.2 for finding the Whole:

$$Whole = Part \, / \, Percent$$

or

$$Whole = \frac{Part}{Percent}$$

Plugging in our values, we get

$$Whole = \frac{15084}{0.012}$$

And the result of the division is 1,257,000. So the salesman's total sales for the month must've been $1,257,000.

You-Try-It 2

a) A new hire at a car dealership is paid a base salary of $1,000 a month plus a 1.25% commission on total car sales for the month. What was the salesman's pay for the month in which he sold 3 cars for a total amount of $75,680?

b) A Wall Street stockbroker is paid purely on commission. If he brought in a total of $4,565,200 in revenue for the month and his pay was $194,021, what is his commission rate?

5.6.3 Percent Increase or Decrease

When quantities change over time it's often more useful to represent the change as a percent rather than simply "by how much." This is often the case where the amounts under consideration vary greatly from place to place. A good example of this is the cost of real estate.

In a typical city in the Midwest such as Fort Wayne, Indiana, the median price of a single-family house may fall in the range of $75,000 to $150,000. In a major city, one that is considered a desirable place to live like San Francisco, the median price may exceed $1,000,000.

Clearly, price *changes* in these two very different markets will also be spectacularly different.

A change in the median home price in SF from $1,000,000 to $1,060,000 is large compared to a median price change of $75,000 to $90,000 in Fort Wayne: $60,000 versus $15,000. But if we just compared these numbers, noting that the increase in SF was 4 times what it was in Ft. Wayne, we would be missing the really important information hidden in these numbers.

The really important information is how much the price changed *compared to what it was before*. The median price change in Fort Wayne was $15,000 compared to $75,000. A common way to compare that is as a *ratio*: 15,000 *to* 75,000.

Ratios are also written like this: 15,000 : 75,000; and also simply as a fraction, $\frac{15000}{75000}$.

Normally we reduce fractions, and we usually do that with ratios written as a fraction. The fraction $\frac{15000}{75000}$ reduces to $\frac{1}{5}$. This fraction represents how much the price changed compared to what it was. The median home price in Ft. Wayne increased in price by one-fifth *of what it was*.

Let's do the same for the increase in the median home price in San Francisco: an increase of $60,000 from the earlier price of $1,000,000. As a fraction, this ratio is $\frac{60000}{1000000}$, which reduces to $\frac{3}{50}$.

Now what we'd like to do is compare these two fractions, $\frac{1}{5}$ and $\frac{3}{50}$. It may not be immediately obvious to you how these compare in size because they are *unlike* fractions (different denominators; refer to 3.2.2).

One of the advantages of using percents is that it makes it much easier to compare two quantities such as $\frac{1}{5}$ and $\frac{3}{50}$. To convert each of these fractions to a percent we multiply by 100. Go back to section 5.2.2 to review this if necessary.

$$\frac{1}{5} = 20\% \quad \text{and} \quad \frac{3}{50} = 6\%$$

We see now that the *relative* increase in the median home price in Fort Wayne was significantly more than the relative increase in the median home price in San Francisco—20% compared to 6%, more than 3 times as much—despite the fact that the actual median increased *amount* in SF ($60,000) was fully 4 times what it was in Fort Wayne ($15,000)!

This is the important information that comes out of these statistics on the increase in the median home prices in SF and Fort Wayne. This quantity, 20% in Fort Wayne and 6% in SF, is what we call the *Percent Increase*.

We got this Percent Increase by forming the ratio of the price change to the original price—the fraction of the change in the price over the original price—and then multiplying that number by 100 to turn it into a percent. So,

$$Percent\ Increase = \frac{Change\ in\ amount}{Original\ amount} \times 100\%$$

The *Original amount* is always the first in time.

Let's simplify the words a bit to make a useable and more easily memorizable formula:

$$\%\ Increase = \frac{Change}{Original} \times 100\%$$

Before we go further and look at *Percent Decrease*, let's do a couple of examples of Percent Increase.

Example. In 1964 the average MLB salary was $15,000. By 2010 it had grown to $3,300,000. Find the Percent Increase.

Step 1.	Our formula, $\% \text{ Increase} = \frac{Change}{Original} \times 100\%$ requires that we first calculate the *Change*.	*Change =* 2010 salary – 1964 salary *Change* = 3,300,000 – 15,000 *Change* = 3,285,000
Step 2.	The original amount is the first in time, $15,000, the salary in 1964. Do the calculation $\frac{Change}{Original}$ on a calculator.	$\frac{3,285,000}{15,000} = 219$
Step 3.	Multiply the result by 100 to turn it into a percent by adding 2 zeros.	219 x 100% = 21,900% This is our Percent Increase from 1964 to 2010. Yes, it is HUGE.

You-Try-It 3

a) In 1964 the average cost of a new car was $3,500. In 2010 the average cost of a new car was $29,225. What is the Percent Increase?

b) The median U.S. home price in 1964 was \$18,000. In 2010 the median home price was \$221,760. Calculate the Percent Increase.

Percent Decrease

Some good news! *Percent Decrease* works almost exactly the same as Percent Increase. The formula is identical. We use Percent Decrease to measure quantities that have gone down over some period of time.[5]

Here is our formula:

$$Percent\ Decrease = \frac{Change\ in\ amount}{Original\ amount} \times 100\%$$

As before with *Percent Increase*, the *Original amount* is always the first in time.

Note when calculating the *Change* in a Percent Decrease, we subtract the *later* amount from the *earlier* amount, since the quantity is decreasing. This is the opposite of what we do when calculating the *Change* in a Percent Increase, where we subtract the *earlier* quantity *from* the *later*.[6]

And, as before with Percent Increase, let's simplify the words a bit for our final formula:

$$\%\ Decrease = \frac{Change}{Original} \times 100\%$$

------

[5] There are no negative numbers involved in Percent Decrease—which is a good thing for us since we are not delving into negative numbers in this book.

[6] This guarantees that both Percent Increase and Percent Decrease will always be positive quantities. A negative Percent Increase would simply be a Percent Decrease, and vice-versa.

Example. The median US home price in May 2007 was $245,000. By May 2015 it had decreased to $178,850. What was the Percent Decrease over this 8-year time span?

Step 1.	Our formula, $\% \; Decrease = \dfrac{Change}{Original}$ **x** 100% requires that we first calculate the *Change*.	*Change* = 2007 price − 2015 price *Change* = 245,000 − 178,850 *Change* = 66,150
Step 2.	The original amount is the first in time, $245,000, the price in 2007. Do the calculation $\dfrac{Change}{Original}$ on a calculator.	$\dfrac{66{,}150}{245{,}000} = 0.27$
Step 3.	Multiply the result by 100 to turn it into a percent by moving the decimal point 2 places.	0.27 **x** 100% = 27% This is our Percent Decrease from 2007 to 2015.

You-Try-It 4

In Sept. 2011 the price of an ounce of gold reached a high of $1,895. In July 2015 it fell to $1,137. What was the Percent Decrease?

5.6.4 Finding a percent of a percent

Consider this scenario. At a college, 64% of the students are female. Twenty-five percent of these female students commute to school by car. A company produces a product or service for female college students who drive to school and wants to know how many of them there are at this particular college.

If the total student population is 8,125, how many students does the company need to target for their ad campaign?

At the simplest level, this is a problem where we seek to find a part of a whole. The whole is the entire student population (8,125). But the *Part* is actually a part of a part. We want to know the part of the female students (64% of the total) that commute by car (25%). What we need to calculate is 25% of 64% of 8,125.

When finding the Part in a basic percent problem, we have to multiply the Whole by the Percent. This follows from the formula from 5.5.1:

$$Part = Percent \ \text{x} \ Whole$$

We *could* do this problem in two separate steps:

1. Find the part of students that are female.
2. Find the part of the female students that commute by car.

Here's how this would work:

Step 1.	Find the number of female students, 64% of the total students.	*Female students =* 64% of 8,125 *Female students* = 0.64 x 8125 *Female students* = 5,200
Step 2.	Find the number of female students that drive to school, which is 25% of the female students.	*Female commuters =* 25% of 5,200 *Female commuters =* 0.25 x 5,200 *Female commuters =* 1,300

But there is simpler way to do this! In math the word "of" translates to multiplication. Here we are looking for a part *of* a part. The parts are identified as percents.

To find 25% of 64% (or 64% of 25%, it doesn't matter), we multiply the percents, ***after changing them to decimal numbers:***[7]

$$25\% \text{ x } 64\%$$

$$0.25 \text{ x } 0.64$$

$$0.16$$

which corresponds to16%.

Now we see that what we have to find is 16% of the total students to find out how many of them are female and commute by car:

$$\textit{Female commuters} = 16\% \text{ of } 8125$$

$$\textit{Female commuters} = 0.16 \text{ x } 8125$$

$$\textit{Female commuters} = 1{,}300$$

Let's now put this together and re-do the Step-by-step analysis from above:

Example. Of a total of 8,125 students at a acollege, 64% are female. Of these, 25% commute by car. How many female students commute by car?

Step 1.	The problem is to find a part of a whole given a percent. The Whole is 8,125. The Percent has to be found by taking a percent of a percent.	*Part = Percent* x *Whole* *Percent* = 25% of 64%

---

[7] When taking a percent of a percent you must change the percents to numbers (typically, decimals) before multiplying them. If you don't do this you will produce a percent that is 100 times too big. If we made this mistake in this problem, 25% x 64% = 1600%, and then took 1600% of 8,092, we would arrive at the nonsensical result that of the 8,125 total student, 130,000 are female who commute by car. This illustrates why we must convert percents to real numbers before performing any actual calculations with them.

Step 2.	Change the percents to decimals and multiply them.	25% of 64%
		0.25 x 0.64
		0.16
Step 3.	Take *this* percent of the Whole.	*Part* = 16% of *Whole*
		Female commuters =
		0.16 x 8125
		Female commuters = 1,300

You-Try-It 5

Thirty-nine percent of registered voters are Republican or 'lean Republican.' Of those, 59% are men. If there are 146,310,000 total registered voters, how many are Republican or Republican-leaning men?

Percents of percents—a last look

Colleges and universities are often exempt from property tax. Some people question the fairness of this. Some Ivy League universities have assets in the $billions. In Massachusetts, the colleges came to an agreement with the state that 25% of the normal business property tax rate is a fair amount that they should volunteer to pay. A news reporter made the following statement, "Some colleges are paying only 13 or 19 percent of what they should be paying."

Without further clarification, it is not clear if these colleges are paying 13 or 19 %, respectively, *of the full business property tax rate*, or if they are paying 13 or 19 percent *of the agreed-upon 25%* of the full business property tax rate. There is a big difference in these amounts. It is in situations like this that clarity—a clear understanding of the math—is necessary to understand what is going on to avoid drawing faulty conclusions.

Let's compare the differences.

Let's name *College A* as the one paying 13 percent, and *College B* as the one paying 19 percent.

If these amounts are the actual percents of the *full* property tax they would pay if they were a typical business, then College A is paying 13% and College B is paying 19% of the rate a neighborhood gas station would pay. But if these are percents *of the 25%* that the colleges agreed to pay, they are far less.

College A:

Percent of percent	13% of 25%
Change to decimal	0.13 x 0.25
Multiply	0.0325
Change to percent	3.25%

Is College A paying 13% of the normal business property tax rate, or 3.25%? ...big difference.

Similarly with College B:

Percent of percent	19% of 25%
Change to decimal	0.19 x 0.25
Multiply	0.0475
Change to percent	4.75%

Is College B paying a property tax amount at least close to the arranged compromise, 19%; or is it paying something at the miniscule rate of less than 5%, less than one-fifth of the agreed upon compromise of 25%?

This example illustrates how sometimes we are not concerned about the actual amounts, but just the (final) percent itself. The difference in the value of the property and assets of Harvard University is far, far more than the value of the property of one of the State University campuses in another part of the state. But it is the property tax *rate* that matters. And those rates are usually given as a percent.

In addressing an issue like this one, where we are concerned with a percent of a percent, we need to be sure that we know how to calculate it properly. Before we leave this section, let's make sure that we can compute percents of percents correctly, without actually computing a corresponding amount, just like in this example we never mentioned the actual value of the properties from which to compute the actual tax bill of the colleges.

You-Try-It 6

Compute, as a percent:

a) Thirty percent of 80%   b) 2.5 % of 50%

c) 75% of 3%   d) 175% of 30%

You-Try-It answers

1a. $28.26	1b. 6%	1c. $160,000	
2a. $1,946	2b. 4.25%	3a. 735%	3b. 1,132%
4. 40%	5. 33,665,931		
6a. 24%	6b. 1.25%	6c. 2.25%	6d. 52.5%

Exercises 5.6

Tax

1. If the sales tax rate is 6.25%, what is the tax on a TV that costs $400?

2. The city assesses the value of your house at $165,000. If the yearly property tax is $2,640, what is the property tax rate?

3. If the import tax on a luxury imported car is 7.5%, what was the cost of a car that was taxed $8,430?

Commission

4. A real estate agent receives a 2.5% commission on a house sale. If her commission on the sale came to $3,550, what did the house sell for?

5. A car salesman's monthly pay is partly based on commission. He receives $750 base monthly pay plus 2% commission on his total monthly sales. If he only sold two cars in the month, for $18,500 and $22,300, what was his total monthly pay for the month?

6. A stockbroker's pay is based purely on commission. If he was paid $4,025 for $161,000 revenue, what is the commission rate, as a percent?

Percent Increase/Decrease

7. In 1990 a copy of Avengers #1 in Fine condition was valued at $550. If the same comic book in the same condition recently changed hands for $3,025, what was the percent increase in the value?

8. In 1960 there were 48,500,000 cigarette smokers in the U.S. By 2015 the number had shrunk to 42,100,000. Find the percent decrease. Round to the nearest whole percent.

Percent of percent

9. What is 12% of 80% (as a percent).

10. Find 160% of 45% (as a percent).

A. Answers to Exercises

Chapter 2

2.2

1. 89	2. 143	3. 130	4. 166
5. 239	6. 595	7. 95	8. 86
9. 93	10. 205		

2.3

1. 74	2. 46	3. 94	4. 47
5. 1,109	6. 68,909	7. 3,819	8. 799
9. 64	10. 28	11. 67	12. 36

2.4

1. 238	2. 348	3. 2,024	4. 688
5. 888	6. 6,804		

2.5

1. 54	2. 39	3. $84\frac{1}{2}$	4. $27\frac{7}{9}$
5. 35	6. 32	7. 11	8. 17
9. $25\frac{5}{12}$	10. $18\frac{1}{2}$		

Chapter 3

3.1

1. $\frac{6}{8}$; $\frac{9}{12}$; $\frac{12}{16}$　　2. $\frac{4}{10}$; $\frac{6}{15}$; $\frac{8}{20}$　　3. $\frac{6}{4}$; $\frac{9}{6}$; $\frac{12}{8}$　　4. $\frac{14}{8}$; $\frac{21}{12}$; $\frac{28}{16}$

5. $\frac{4}{5}$　　6. $\frac{1}{6}$　　7. $\frac{4}{3}$　　8. $\frac{3}{7}$

9. $\frac{7}{10}$　　10. $\frac{4}{75}$　　11. 8　　12. 7

13. 0　　14. 1　　15. $\frac{8}{3}$　　16. $\frac{23}{4}$

17. $\frac{43}{5}$　　18. $\frac{77}{6}$　　19. $4\frac{1}{3}$　　20. $3\frac{1}{9}$

21. $6\frac{3}{7}$　　22. $2\frac{3}{13}$　　23. $\frac{4}{3}$　　24. 9

25. $\frac{1}{7}$　　26. Undefined　　27. $\frac{25}{14}$　　28. 1

3.2

1. $\frac{7}{9}$　　2. $\frac{5}{7}$　　3. 3　　4. $\frac{10}{7}$ or $1\frac{3}{7}$

5. 18　　6. 60　　7. 30　　8. 40

9. 30　　10. 180　　11. 18　　12. 24

13. 60　　14. 30　　15. $\frac{5}{4}$ or $1\frac{1}{4}$　　16. $\frac{3}{2}$ or $1\frac{1}{2}$

17. $\frac{5}{6}$　　18. $\frac{14}{15}$　　19. $\frac{13}{10}$ or $1\frac{3}{10}$　　20. $\frac{97}{60}$ or $1\frac{37}{60}$

21. $8\frac{3}{7}$　　22. $4\frac{11}{12}$　　23. $10\frac{11}{15}$　　24. $15\frac{5}{8}$

25. $10\frac{1}{6}$　　26. $12\frac{41}{60}$

3.3

1. $\frac{1}{3}$

2. $\frac{7}{11}$

3. $\frac{1}{4}$

4. $\frac{2}{9}$

5. $\frac{2}{5}$

6. $2\frac{4}{7}$

7. $1\frac{1}{2}$

8. $3\frac{1}{2}$

9. $1\frac{3}{4}$

10. $3\frac{1}{6}$

11. $\frac{9}{10}$

12. $1\frac{31}{48}$

3.4

1. $\frac{3}{10}$

2. $\frac{9}{32}$

3. 1

4. $\frac{7}{10}$

5. $\frac{4}{3}$

6. $\frac{3}{5}$

7. $\frac{48}{5}$ or $9\frac{3}{5}$

8. $\frac{75}{11}$ or $6\frac{9}{11}$

9. $\frac{28}{3}$ or $9\frac{1}{3}$

10. $\frac{189}{5}$ or $37\frac{4}{5}$

3.5

1. 4

2. $\frac{3}{8}$

3. $\frac{5}{9}$

4. 5

5. $\frac{3}{2}$

6. $\frac{8}{15}$

7. $\frac{1}{2}$

8. $\frac{21}{4}$ or $5\frac{1}{4}$

9. $\frac{2}{15}$

10. $\frac{23}{10}$

Chapter 4

4.1

1. 5.1

2. 5.11

3. 5.1096

4. 5.110

5. 5

6. 10

7. $\frac{3}{25}$

8. $\frac{3}{8}$

9. $\frac{7}{500}$

10. $\frac{9}{100}$

11. $\frac{5}{4}$

12. $\frac{21}{20}$

13. $\frac{102}{25}$

14. $\frac{101}{10}$

15. $1\frac{2}{5}$

16. $5\frac{1}{20}$

17. $3\frac{1}{8}$

18. $47\frac{13}{200}$

4.2

1. 23.7

2. 9.78

3. 20.59

4. 20.282

5. 1.743

6. 39.0613

7. 4.21

8. 2.37

9. 1.71

10. 2.946

11. 2.653

12. 3.6224

4.3

1. 36.9

2. 115.24

3. 0.441

4. 6.7824

5. 128.0686

6. 84.07959

7. 35.2

8. 14,065

9. 854

10. 68.09

4.4

1. 8.5

2. 22.6

3. 17.8

4. 23.5

5. 0.316

6. 8.204

7. 5.0738

8. 0.014903

9. 0.25

10. 1.1

11. 0.125

12. 0.6

13. 0.625

14. 2.5

15. $0.\overline{3}$

16. $0.8\overline{3}$

17. $0.\overline{4}$

18. $0.\overline{142857}$

Chapter 5

5.2

1. 0%

2. 1,000%

3. 700%

4. 2,500%

5. 25%

6. 20%

7. 70%

8. 62.5%

9. $68\frac{3}{4}\%$

10. $33\frac{1}{3}\%$

11. $83\frac{1}{3}\%$

12. $44\frac{4}{9}\%$

13. $433\frac{1}{3}\%$ or $433.\overline{3}\%$

14. 340%

15. $222\frac{2}{9}\%$ or $222.\overline{2}\%$

16. 1,250%

17. 450%

18. 340%

19. $266\frac{2}{3}\%$ or $266.\overline{6}\%$

20. 625%

21. 75%

22. 50%

23. 12.5% or $12\frac{1}{2}\%$

24. 6.25% or $6\frac{1}{4}\%$

25. 145%

26. 1,009%

27. ~19%

5.3

1. 0.56

2. 0.05

3. 1.4

4. 20

5. $\frac{3}{4}$

6. $\frac{89}{100}$

7. $\frac{1}{5}$

8. $\frac{3}{50}$

9. $\frac{16}{5}$; $3\frac{1}{5}$

10. $\frac{5}{4}$; $1\frac{1}{4}$

11. $\frac{13}{5}$; $2\frac{3}{5}$

12. $\frac{112}{25}$; $4\frac{12}{25}$

13. 0.722

14. 0.155

15. 1.128

16. 0.00025

17. 0.0012

18. 0.0399

19. 0.055

20. 0.0075

21. 0.0658

22. 0.06375

23. $\frac{11}{200}$

24. $\frac{3}{400}$

25. $\frac{329}{500}$

26. $\frac{51}{800}$

27. $0.04\overline{6}\%$

28. $0.05\overline{1}\%$

29. $\frac{7}{150}$

30. $\frac{23}{450}$

5.4

1. 37	2. 40	3. 39	4. 66
5. 155	6. 2,800	7. 85	8. 400
9. 30%	10. 7%	11. 1,812.5%	12. 1.6%

5.5

1. 8.1	2. $271.25	3. 300	4. 4
5. 80	6. 180	7. 16	8. 35
9. 3.5%	10. 27.5%	11. 2,280%	12. 35%

5.6

1. $25	2. 1.6%	3. $112,400	4. $142,000
5. $1,566	6. 2.5%	7. 450%	8. 13%
9. 9.6%	10. 72%		

Made in the USA
Middletown, DE
07 September 2019